PREFACE

I wish to express my appreciation to
Doctor G Steiner, Doctor J R. Christie and
Miss E M Buhrer, of the Division of
Nematology. U S Bureau of Plant Indus-
try, for their helpful suggestions concerning
certain subjects and criticisms of the
chapters I have written wholly or in part
To Dr H J Van Cleave of the Department
of Zoology, University of Illinois, gratitude
is due for his criticisms of the chapter on
"Nemic Relationships", as was to be expected he did not wholly agree with the
treatment of this controversial subject Mr
Jonas Bassen of the U S Bureau of En-
tomology and Plant Quarantine has aided
through the translation of Russian articles
and Miss Dorothy Bero has checked the
bibliographies

Mr G A Grille of the Spencer Lens
Company, Washington, D C, kindly sup-
plied the photograph of Dr. Hall

Section I, Part III includes a chapter
by Doctor Reed O Christenson with contributions by Dr F G Wallace, Mr Leon
Jacobs and Mrs M B Chitwood Section
II, Part I contains chapters by Dr A C
Walton and Mrs M B Chitwood, they assume full responsibility for the contents of
their chapters unless otherwise noted I
wish to express my appreciation for the
fine way I feel they have handled their
subjects I assume responsibility for the
nomenclature used.

B G C.

CHAPTER I

GAMETOGENESIS

A. C. WALTON KNOX COLLEGE, GALESBURG, ILL

The history of the formation of the germ cells among nematodes is so closely bound up with the processes of meiosis and fertilization that consideration of any one of these phenomena involves a discussion of all three

The process of meiosis, or reduction division, was first announced by Van Beneden in 1883 in his report of studies on the egg and spermatozoon of *Parascaris equorum* (*Ascaris megalocephala*) and the fact that the gametes contained only one-half the number of chromosomes found in the body cells, equally divided as to origin from each parent, is one of the most fundamental concepts of the fields of Evolution and Heredity The realization that *Parascaris* germ cells were large, easily obtained, and very simple in their nuclear organization, led to their use as study material in the rapid advances of Cytology during the last decade of the nineteenth century

The germinal cells of nematodes are differentiated during the very early cleavage divisions of the zygote and furnish a very clean history of germ-cell isolation, especially in those forms which show the "diminution" phenomenon Ignoring for the time this peculiar process, the mitotic activity of the somatic and of the germinal cells has afforded a fruitful source for cytological investigations It was from the study of *Parascaris* (*Ascaris megalocephala*) that Van Beneden (1883), Boveri (1887, 1888, 1890), Herla (1893), and Zoja (1896) laid the foundation work that established the doctrine of the genetic continuity of chromosomes, not only as to material, but also as to individual size and shape The same material allowed Boveri (1909), Bonnevie (1908, 1912), and Vejdovsky (1912) to work out the structure of the individual chromosomes, a result that later workers on other materials have largely substantiated as to the main interpretations (For a review of the literature up to 1923 see Walton, 1924)

As a result of these and other studies on nematode materials, the process of somatic mitosis seems to fall in line with the general system as follows The reticulum of the nucleus becomes organized into a number of fine chromidial threads (Brauer, 1893) during the early prophase, these undergo an accurate longitudinal splitting, shorten and thicken, and take their places as individual chromosomes in the equatorial plate at the end of the prophase The metaphase proper is practically absent, as splitting occurs early in the prophase During the anaphase the chromosomes separate along the line of longitudinal splitting and pass to the two poles of the achromatic spindle During the telophase each group of chromosomes becomes transformed into a new nuclear reticulum in which the individual chromosomes may lose their visible outlines, but not their actual identity through vacuolization (Van Beneden, 1883, 1887), branching (Rabl, 1889, Boveri, 1887), or chromomema formation (Vejdovsky, 1912) The somatic number of chromosomes remains constant although they are divided equationally at each division and, since they are all descendants of the chromosomes of the zygote nucleus, the chromatic material of every germ cell and of every body cell is directly derived from that which was brought into the zygote nucleus by the egg and sperm nuclei of the preceding generation, a fact of enormous importance in the study of heredity and development

The achromatic as well as the chromatic elements of the cell have been studied carefully in nematode material Van Beneden (1887) and Boveri (1887) established the

thesis that the centrosome is a permanent and genetically individual cell structure Although usually regarded as extra-nuclear in position, it is reported as of intra-nuclear origin in *P equorum* var *univalens* (Brauer, 1883) and in *P c vni bivalens* (Sturdivant, 1931) In spite of much criticism, modern workers in the same field have substantiated this conclusion, at least as to cells of *Parascaris equorum* (Togg, 1931, Sturdivant, 1934), although the exact nature of the structure is still unknown The centriole divides (Boveri, 1900, Sturdivant, 1934) before any other visible evidence of mitosis appears, and migrates to opposite sides of the nucleus to form the poles of the next spindle figure The spindle proper (first seen in nematode materials by Auerbach, 1874), the mitome ring, and the astral rays appear to be composed of granules and fibers which probably are the result of chemical fixation of what in the living cells, are delimited currents of nuclear material in reaction with certain cytoplasmic elements which center at the centrosomal points, and are not fibers of actual material identity as stated by Boveri (1888) The fibers appear before the nuclear membrane disappears, and their extraor intra-nuclear origin may depend upon the differential permeability of the membrane, streaming may first begin either in the cytoplasm or the karyoplasm, depending upon the physiological condition of the two substances

Cytokinesis, as opposed to karyokinesis, is usually accomplished by a process of constrictive furrowing caused by differential surface tension and surface streaming phenomena (Spek, 1918 and 1920 in *Rhabditis pellio* and *R dolichura*) which seem to depend upon the changes in the permeability of the cell membrane These phenomena seem to be correlated with karyokinesis through the medium of the achromatic spindle

Meiosis, as a phenomenon, accomplishes the "reduction of chromosomes" in that it affords an opportunity for the numerical reduction of the constant somatic complement of chromosomes (the diploid number) to the gametic (haploid number) and also separates the members of each pair of homologous chromosomes present in the somatic complex In such a process, two forms of chromosome division occur, (a) separation equationally of split chromosomes, and (b) disjunction of homologous (paired) structures As in most animals, meiosis occurs in connection with gametogenesis among the nematodes. In the male those cells (spermatogonia) destined to give rise to the spermatozoon undergo a series of ordinary equational divisions until a certain definite number is reached The last generation of these cells undergoes a growth period during which the homologous male- and female- derived chromosomes are paired The resultant cells (the primary spermatocytes) have the *haploid* number of chromosome *pairs* Two successive meiotic divisions, one disjunctive and the other equational, follow without complete nuclear reorganization during the interphase The first division gives rise to two secondary spermatocytes and the second divides the two secondary spermatocytes into four spermatids Normally each spermatid metamorphoses into a spermatozoon, giving four spermatozoa (male gametes) as the end result of the two meiotic divisions In certain of the free living nematodes Cobb (1925, 1928) reports the intercalation of a number of equational divisions of the spermatid before the ultimate differentiation of the spermatozoa (*spermules*). In *Spirina parasitifera*[*] each spermatid eventually gives rise to one hundred and twenty-eight spermatozoa, the final differentiation occurring only after the spermatogenous tissue reaches the oviducts of the female. Certain arachnids (Warren, 1930), which produce two to four spermatozoa from each spermatid, show an approach to this condition. A somewhat similar phenomenon is also reported from certain snails.

In the female a similar program is followed to a great extent, except that the meiotic divisions frequently occur only after the spermatozoon has entered the primary oocyte (Boveri 1887; Sala, 1895; and modern observers). The spindle of the first meiotic division is always eccentric in position and the resultant cells are extremely unequal as to cytoplasmic content. The first polar body is separated from the large secondary oocyte by this division. The second division similarly forms a single large functional ovum and a second small polar cell. The first polar body occasionally divides into two equal cells. None of the polar cells are functional as far as is known among nematodes, although cases of entrance of the sperm into such cells have been noted.

As stated above, the significance of meiosis lies in the reduction to n/2 of the chromosomes which have undergone synapsis during the formation of *bivalent* chromosomes (homologous pairs). This synapsis is now regarded as being always "side by side" (para-synapsis) among the nematodes. If such members of homologous pairs are compound chromosomes, their synapsis may give rise to four-parted *bivalents* during the prophase of the first meiotic division. This occurs during the growth period following the last gonial division of each germ cell of either sex. These bivalents are separated during one of the two following divisions and hence that division is reductional since it separates (disjunction process) homologous structures. The other division, being equational, means that the four resulting nuclei each have a haploid set made up of one chromosome of each kind. In many nematodes the prophase chromosomes of the first division show both the plane of synapsis and that of a future longitudinal splitting, making "tetrad" chromosomes (if the chromosomes are compound they may then form "di-tetrads", or, as in *Spirina parasitifera*, they may form 56-parted bodies). When the first division separates homologous pairs, it is termed "prereductional"; when it is the second division which causes disjunction, the process is termed "postreductional". Most nematodes show "prereduction". If the original bivalent chromosomes were "tetrads" the resultant chromosomes are "monads"; if "di-tetrads", they become "dyads" in the mature germ cells and in the polar cells.

During the formation of the prophase chromosomes the stages known as leptotène, zygotène, pachytène, diplotène and strepsitène are poorly differentiated except possibly in the races of *Parascaris equorum* (Van Beneden, 1887; Boveri, 1888; Griggs, 1906; Bonnevie, 1908, 1912); in most cases the chromosomes behave as quite solid units derived from a very early stage from a segmented spireme thread (Vejdovsky, 1912; Walton, 1918, 1924; Sturdivant, 1934).

During the process of spermatogenesis extra-nuclear bodies such as the centrioles (?), chondriosomes (Meves, 1911; Held, 1912, 1916; Hirschler, 1913; Romeis, 1912 Sturdivant, 1931, 1934), "yolk granules" (Sturdivant, 1934; Wildman, 1912; Walton, 1916a), and Golgi bodies (Sturdivant, 1934) are more or less evenly distributed so that each spermatid receives its complement of each of these elements in addition to the haploid number of chromosomes. The "yolk granules" are thought to be largely composed of glycogen and to be low in protein and lipoids (Kemnitz, 1913), although Bowen (1925) believes that further analyses are needed. These "yolk granules", or refringent globules, are apparently derived through the activity of the Golgi bodies, and therefore are pro-acrosomal in nature. The contained bodies in the center of each globule disappear during the spermatid metamorphosis and are perhaps to be regarded as temporary indications of precocious acrosomal granules, structures quite characteristic of insect spermatozoa (Sturdivant, 1934). They are not mitochondrial in nature

as earlier reported. The refringent globules eventually fuse to form the "refringent body" of the mature spermatozoön, which thus contains a structure homologous with the acrosome of other types of spermatozoön (Bowen, 1925). The Golgi remnants are cast off during the cytoplasmic reduction and cytophore formation of the maturing spermatid (Sturdivant, 1934).

The spermatozoa of nematodes are described as non-flagellated, frequently amoeboid cells, containing a considerable amount of stored material in the "refringent body", or acrosome. This type of spermatozoön is usually regarded as a simple modification of the fundamental structural plan of a flagellate sperm and has arisen secondarily during the evolution of this phylum. The fact that the acrosome is not always at the morphologically anterior end of the spermatozoön is not of particular significance. Certain of the acrosomal bodies are hollow (*Nemotospira turgida*), and this may very doubtfully represent the position of an axial tail filament in this pseudo-flagellate form. No other evidence concerning any axial filament is apparently available for the nematodes. *Passalurus ambiguus* (*Oxyuris ambigua*) has spermatozoa that may almost be considered as flagellate (Meves, 1911; Bowen, 1925). Recently Chitwood (1931) has described the spermatozoa from *Trilobus longus* which seem to be of truly flagellate form. This is to be expected since *Trilobus* is very close to the hypothetical ancestral nematode form which is believed to have possessed a typically flagellate type of sperm. *Passalurus* (*Oxyuris*) and *Trilobus* have the acrosomal body at the morphologic-

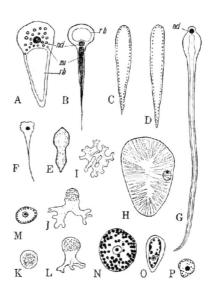

Fig. 147.

Nemic spermatozoa. A.—*Parascaris equorum*; B.—*Passalurus ambiguus*; C.—*Anticoma pellucida*; D.—*A. eberthi*; E.—*Trichosomoides crassicauda*; F.—*Tetradonema plicata*; G.—*Trilobus longus*; H.—*Dorylaimopsis metatypicus*; I-L.—*Rhabditis strongyloides* (I, J, L, amoeboid stage, various views; K, resting stage); M.—*Paracanthonchus viviparus*; N.—*Halichoanolaimus robustus*; O.—*Tripyla papillata*; P.—*Aracholaimus spinosus*. B, after Meves, 1920, Arch. Mikr. Anat. v. 54; C, after de Man, 1886, Nordsee-Nematoden, Leipzig; D, after de Man, 1889, Mem. Soc. Zool. France, v. 2; F, after Cobb, 1919, J. Parasit., v. 5; G-P, original, Chitwood; A, E, original, Walton.

[*] Chitwood has re-examined this form and reports that the above observation was based on a misinterpretation of the structures present. (see page 125 for his explanation).

206

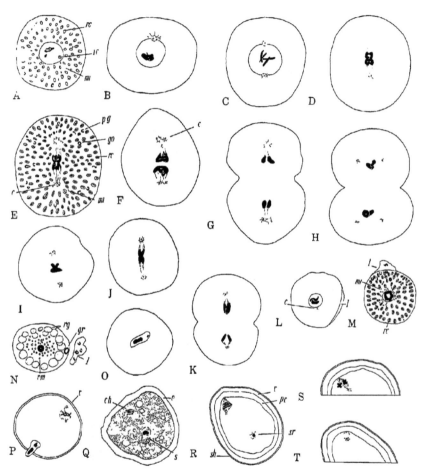

Fig 148

Gametogenesis A —*Parascaris equorum*, Spermatogonium (showing intranuclear centrosome mitochondria refringent corpuscles with polar bodies and nuclear contents) B —*Parascaris equorum* Early prophase of 1st spermatocyte (extrusion of intranuclear centrosome) C —*Parascaris equorum* Late prophase of 1st spermatocyte D —*Parascaris equorum* Metaphase of 1st spermatocyte E —*Parascaris equorum* Late metaphase of 1st spermatocyte (centrosomes dividing) F —*Parascaris equorum*, Anaphase of 1st spermatocyte G —*Parascaris equorum*, Early telophase of 1st spermatocyte H —*Parascaris equorum*, Telophase of 1st spermatocyte (centrosomes divided) I —*Parascaris equorum* Late prophase of IInd spermatocyte J —*Parascaris equorum* Metaphase of IInd spermatocyte K —*Parascaris equorum* Telophase of IInd spermatocyte L —*Parascaris equorum* Early spermatid (cytoplasmic lobe forming) M —*Parascaris equorum* Later spermatid (cytoplasmic structures indicated) N —*Parascaris equorum* Spermatid (cytoplasmic reduction completed) O —*Parascaris equorum* Oogonium of last generation (intranuclear centrosome) P —*Parascaris equorum* Late prophase of 1st oocyte (penetration of spermatozoon) Q —*Parascaris equorum*, Prophase of 1st oocyte (sperm with divided centrosome) R —*Parascaris equorum* Late prophase of 1st oocyte S —*Parascaris equorum*, Metaphase of 1st oocyte (tetrad formation) T —*Parascaris equorum* Telophase of 1st oocyte All drawings original

207

ally anterior end of the spermatozoon. It is perhaps to be expected that in *Parascaris*, and in related genera where chromosomal behavior as well as other criteria point to a high degree of specialization, the acrosome likewise would tend to vary from the normal, as perhaps is shown by its unusual position behind the nucleus. Some nematodes show distinct polymorphism in sperm size, a condition believed to be correlated with the difference in chromosomal numbers between the "male-producing" and the "female-producing" male gametes (Goodrich, 1916; Meves, 1903; Mulsow, 1911). This chromosome variation is most clearly demonstrable in species in which there is a complex type of "X" chromosome, often involving a large number of chromatin elements (Walton, 1924).

The completion of the germ cycle involves the process of syngamy by which the union of gamete nuclei and the restoration of the diploid number of chromosomes is accomplished. Syngamy in nematodes is complicated by the fact that the maturation of the egg and fertilization proceed simultaneously, the spermatozoon frequently entering the egg during the prophase of the first meiotic division.

The entire spermatozoon, at least among those that are amoeboid in form, enters the egg and immediately a thick fertilization membrane forms, appearing first near the point of entrance and finally enclosing the entire egg. The reticulated male pronucleus gradually forms from the condensed spermatozoon nucleus, the mitochondrial elements slowly fade into the egg cytoplasm as the male cell wall disappears, and the remnant of the mass of acrosomal material eventually loses its separate identity.

In the case of many nematodes (*Rhabditis bufonis*, *R. runae*, *Rhabditis terricola*, *Syphacia obvelata* (*Oxyuris obvelata*), *Turbatrix aceti*) the shell membrane is reported as being applied to the egg before the entrance of the spermatozoon†. In such cases a micropylar opening has been described, usually at the end of the egg which was originally attached to the rhachis, and opposite to the pole at which the polar cells are normally extruded. No such structure is necessary in the forms in which sperm penetration precedes egg-shell formation. A structure resembling a micropyle has been described in forms which have the egg shell formed after sperm entrance has occurred. In *Ascaridia galli* (*A. lineata*) Ackert (1931) has shown that this is not a true micropyle, and perhaps similar micro-dissection studies might necessitate the revision of the descriptions of the presence of a micropyle in several forms. If a true micropyle is present, it seems obvious that the sperm entrance is fixed at what may be regarded as the vegetative pole of the egg, since many observers have determined that the first polar cell is eliminated at a point opposite the entrance path of the spermatozoon. Probably the same statement holds for those forms in which sperm entrance precedes shell formation, inasmuch as sperm entrance and first polar cell positions are directly opposite in most nematode eggs, and the point of sperm entrance in *Parascaris equorum* has been shown by Schleip (1924) to be at the originally attached end. This problem is tied up with that of the polarity of the egg which is discussed elsewhere.

The two pronuclei come to lie side by side, the first cleavage spindle is established, and division follows. During this process the male and female chromosomes occupy opposite sides of the spindle and it is not until the second cleavage division that the two sets of chromosomes are indistinguishably mixed, although in some cases complete intermingling may be delayed until later in the cleavage phenomenon.

The development of the egg without fertilization (true parthenogenesis) is rare among nematodes although two species of *Rhabditis* (Belar, 1923) have been described as showing only a single maturation division and no reduction in the chromosome number. Krüger (1913) reports that the hermaphroditic *Rhabditis aberrans* (probably a variety of *R. aspera*) produces eggs that are apparently parthenogenetic of the diploid type (one polar cell and no chromosome reduction of the somatic number of 18) although frequently the sperm actually enters the egg but degenerates and fails to enter the cleavage nucleus. In a normally dioecious *Rhabditis pellio* culture, P. Hertwig (1920) found a mutant which produced only one polar cell without reduction, and thus retained the diploid number (14). None of these eggs would develop unless entered by a sperm, but again in no case did the sperm contribute to the cleavage nucleus. These two cases bridge the gap between normal fertilization and normal parthenogenesis.

Many nematode species show a "diminution" phenomenon (Walton, 1918, 1924) in the non "stem-cells" of early cleavage, examples occurring from the second to the sixth division, and then ceasing, as by the sixty-four-cell stage the primordial germ cells are entirely differentiated. The process of "diminution" which involves the elimination of a portion of each of the chromosomes in the nucleus is best known in the embryonic cells of *Parascaris equorum*. In this form the process may begin in the second cleavage of the soma cells although it usually first appears in the third cleavage, and then is found in the division of each new soma cell separated from the "stem" cell until the "germ line" cells are definitely isolated. In *P. equorum* this process is completed during the fifth cleavage. All germ cells retain the undiminished amount of chromatin, while all soma cells have the reduced amount as the result of "diminution". During the prophase of the "diminution division" the chromosomes of the soma cell break up, the center forming a definite number of small chromosomes and the ends several blobs of material. The small chromosomes divide equationally while the larger masses are left behind. The daughter nuclei reorganize without the extruded remnants, which then ultimately degenerate and disappear. The process is quite similiar in other species of nematodes (Meyer, 1895; Bonnevie, 1901; Walton, 1918, 1924) except that there is frequently no increase in number of chromosomes during the process inasmuch as the gametic chromosomes in many species are not as complex as they are in *Parascaris* spp.

† The formation of a shell before fertilization is dubious. See Sect. 1, Part 3, Chapter 12. B. G. C.

Fig. 149.

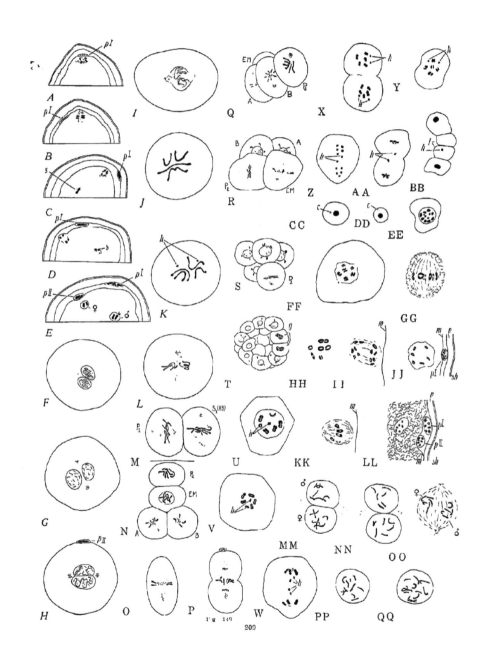

Fig 149

209

The process of diminution is not confined to the nematodes. Members of the Diptera (*Miastor*), Coleoptera (*Dytiscus* and *Colymbetes*), and Lepidoptera (*Lymantria Oigyra, Phragmatobia, Ephestia, Philosamia*, etc.) also show a similar phenomenon. In the nematodes the process always accompanies the localization of the germinal "stem cell" and is confined to those cells which are derived from the "stem cell" but whose descendants become "soma cells." Only the cells which contain the cytoplasmic area destined to become germ cells fail to undergo diminution. Diminution is thus early in somatic history. The same is perhaps true in the case of *Miastor*, where diminution is confined to the last oogonial divisions. It might seem to separate chromatin useful in germ cells, though not in soma cells. In the Coleoptera the process comes late in the germ-line history and does not separate somatic from germinal chromatin. Among the Lepidoptera, diminution occurs after the chromosomes are set free from the nucleus (during metaphase time) and frequently is found only during the maturation divisions of the egg. This variation in the time of occurrence prevents any interpretation of the phenomenon as one of separation of somatic and germinal types of chromatin. The only generally accepted fact common to all cases of diminution is that it is oxychromatin which is lost and basichromatin which is retained. According to Fogg (1930) "The only safe conclusion that now seems admissible is that diminution plays no primary or essential part in differentiating the germ-line from the somatic. It is rather a by-product of conditions existing in the cytoplasm which may vary widely in different species in respect to the time of its occurrence, its modus operandi, and its physiological significance"

Several workers have reported the loss of portions of chromatic material by means other than "diminution." Chief of these methods is through the "cytophore" form-

ation which accompanies the metamorphosis of the spermatid in many animal groups. This phenomenon has been reported for *Cystidicola farionis* (*Aneuracanthus cystidicola*) (Mulsow, 1912), *Toxocara vulpis* (*Belascaris triquetra*) (Marcus, 1908a, Walton, 1918), *Parascaris equorum* (Hertwig, 1890, Mayer, 1908, Sturdivant, 1934), *Ascaris lumbricoides* (Hirschler, 1913), *Rhabdias bufonis* (Boveri, 1911, Schleip, 1911), and *Spirura parasitifera* (Cobb, 1925, 1928)

Among the nematodes the two sexes are normally separate, although a number of hermaphroditic forms are known, particularly among those species which are free-living (Maupas, 1900, Potts, 1910, Cobb, etc) or those which alternate between free-living and parasitic generations (Boveri, 1911, Schleip, 1911). In such cases the parasitic generation is the one showing hermaphroditism. In many of the bisexual forms the males show an "XO" type of sex chromosome (occasionally an "X" complex) and the females an "XX" condition. A similar condition is known in the hermaphroditic generation of *Rhabdias bufonis* and in a single unusual specimen of *P. equorum* var. *bivalens* (Goulhart 1932). In both of these cases the spermatozoa are "XO" and the eggs "XX" in type. The formation of the hermaphroditic generation of *R. bufonis* is probably due to the non-viability of the non "X"-bearing spermatozoon when produced by the free-living males, but in the hermaphroditic generation both types of sperm ("X" and "O") are viable and hence union with the "X"-bearing eggs produces the free-living generation males, "XO", and the females, "XX". The "XY" and "XX" condition is doubtfully reported from several species. The only clear-cut case is one of a multiple "X" and simple "Y", and multiple "XX", from a single species (*Contracaecum incurvum = Ascaris incurva*) by Goodrich (1916). In the great majority of nematodes, the heterochromosome has not been recognized, possibly because, as

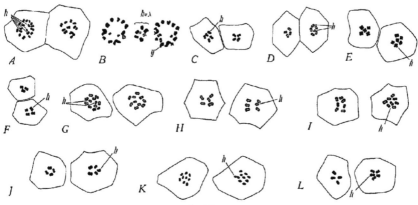

Fig 150

A—*Toxocara canis* IInd spermatocytes (12 & 18 dyad chromosomes X = 6) B—*Contracaecum incurvum*, Anaphase of 1st spermatocyte (18 + lagging X group & 18 + X, X = 8 X = 1) C—*Heterakis papillosa*, IInd spermatocytes (4 & 5 dyad chromosomes X = 1) D—*Heterakis spinosa* IInd spermatocytes (4 & 6 tetrad chromosomes X = 2) E—*Nema torspira turgida* IInd spermatocytes (5 & 6 tetrad chromosomes X = 1) F—*Trichosomates crassicauda* IInd spermatocytes (3 & 4 tetrad chromosomes X = 1) G—*Toxocara vulpis*, IInd spermatocytes (10 & 12 tetrad chromosomes X = 2) H—

Cruzia tentaculata IInd spermatocytes (5 & 6 tetrad chromosomes X = 1) I—*Contracaecum spiculigerum* IInd spermatocytes (7 & 8 tetrad chromosomes X = 1) J—*Mastophorus muris* IInd spermatocytes (4 & 5 dyad chromosomes X = 1) K—*Toxocara cati* (IInd spermatocytes (9 & 9 monad chromosomes heterochromosome X = 1 attached to one autosome) L—*Physaloptera turgida*, IInd spermatocytes (4 & 5 dyad chromosomes X = 1) C after Goodrich 1916 J Exper Zool v 21 others original is so frequently the case in *P. equorum*, it is attached to the end of an autosome and only occasionally is distinct enough for positive identification In most cases the heterochromosome undergoes "pre-reduction" as do the autosomes, but in some cases it shows "postreduction" although the autosomes seem to show the *differential division* as being the first This may point to the primitive condition being actually one of "postreduction" (Edwards 1910, Wilson, 1925, p 757) In many instances the heterochromosome (or heterochromosome complex) either precedes the others or lags behind during one or both of the meiotic divisions and in some cases forms a separate chromatin nucleolus during the interphase stage Where both "X" and "Y" are present, they are separated most frequently at the first division, each undergoing equational splitting at the second In the early Spermatocyte I growth period nuclei the "XY" group is differentiated from a single chromatin nucleolus and the "XY" pair assumes a "tetrad" form, usually asymmetrical because of the small bulk of the "Y" element Even when the "X" is multiple it differentiates from a single nuclear body (Walton, 1916, 1924), just as it does when "X" = 1

The nematodes therefore afford a wide variation of heterochromosome types, varying from a single "X" and no "Y" in *Heterakis dispar* and *Cystidicola farionis* (*Ancyracanthus cystidicola*) to forms like *Toxocara vulpis* (*Belascaris triquetra*) and *Heterakis spumosa* (*Gangulet-erakis spumosa*) with an "X"-complex of two, *Ascaris lumbricoides* with one of five, *Toxocara canis* (*Toxascaris canis*) with one of six, and *Parascaris equorum* with one of eight to nine, and no "Y" and thence to forms such as *Contracaecum incurvum* with an "X"-complex of eight and a single "Y" Peculiarly, no established case of an "XY" pair has been definitely recognized The "X" and "Y"-chromatin may form a single body or a single unit during meiosis, just as frequently the autosomes may conceal their complexity temporarily in single bodies under the same circumstances (*P. equorum*) *

The following chart gives the majority of the examples of the species which have furnished material for the study of nematode gametogenesis In each case the haploid and diploid chromosome numbers are indicated, the somatic number is given, and the form of the chromosomes at each stage (di-tetrad, tetrad, dyad, monad) is noted Wherever the germinal chromosomes are plurivalent, their unit value in terms of the somatic chromosomes is pointed out The nature and number of the heterochromosomes is indicated for each species as far as it is known The presence or absence of the "diminution" process is also stated for such species as have been examined for that phenomenon

---

* Jeffrey and Haertl (1938) have recently questioned the whole subject of sex chromosomes in nematodes in their study on *Ascaris lumbricoides*, *Toxocara* cat *T. canis* and *Ascaris* sp? from a seal They fail to find any evidence of a consistent differential distribution of what might be called X chromatin and show that while certain chromosomes lag behind in each meiotic division these are not necessarily distributed as X and O, or X and Y materials must be Analogous behavior of chromosomes is known to occur in various hybrid forms of plants and animals and is evidence of their mixed ancestry The authors argue that these nematodes and probably all similar forms are likewise hybrids because of the evidence presented by the behavior of their chromosomes particularly during the process of meiosis If the nematodes are hybrids then the uneven distribution of the chromosomes during the maturation divisions is to be expected and is a clue to their hybrid ancestry not a proof of the presence of sex chromosomes as has been the usually accepted interpretation

---

List of Abbreviations

A, Cell formed by the division of the 1st generation Soma cell
B, Cell formed by the division of the 1st generation Soma cell
c Centrosome
ch Chromosomes
EM Cell formed by the division of the 1st generation Stem cell
g, Germ cells
go, Golgi body
r Golgi ring
h Heterochromosome
n, Intranuclear centrosome
l Cytoplasmic lobe
m Cell membrane
nm Mitochondria
ncl Nucleus
P I, 1st polar body
P II 2nd polar body
P1, 1st generation Stem cell
P2, 2nd generation Stem cell
pg Pre-acrosomal granule
Ps Perivitelline space
rb Refringent body or Acrosome
rc Refringent corpuscle
rg, Refringent globule
Sf 1st generation Soma cell
s, Spermatozoon
sh, Egg shell
sr, Sperm remnant
t Fertilization membrane
vm Vacuolated mitochondria
X X-chromosome
Y, Y chromosome
Male sign, Male pronucleus, or of male origin
Female sign, Female pronucleus, or of female origin

Table 7 Chromosome numbers of nematodes

Name	Haploid number of chromosomes	Diploid number of chromosomes	Diminution	Somatic number of chromosomes	Heterochromosome number	Autosome Unit Value
1 *Ascaridia galli* (*Heterakis inflexa*)	5 (tetrads)	9-10 (monads)	undetermined	9-10	X = 1	1
2 *Ascarid* (from dog)	4 (di-tetrads)	8 (dyads)	undetermined	16	?	2
3 *Ascaris anguillae* (*Ascaris labiata*)	?	?	present	?	?	?
4 *Ascaris lumbricoides*	24 (tetrads)	43-48 (monads)	present	43-48	X = 5	1
5 *Camallanus lacustris* (*Cucullanus elegans*)	6 (tetrads)	12 (monads)	undetermined	12	?	1
6 *Contracaecum clavatum* (*Ascaris clavata*)	12 (di-tetrads)	24 (dyads)	undetermined	48	?	2
7 *Contracaecum incurvum* (*Ascaris incurva*)	21 (tetrads)	35-42 (monads)	present	35-42	X = 8, Y = 1	1
8 *Contracaecum spiculigerum* (*Ascaris spiculigera*)	5 (di-tetrads)	9-10 (dyads)	absent	18-20	X = 1	2
9 *Cruzia tentaculata*	6 (di-tetrads)	11-12 (dyads)	undetermined	22-24	X = 1	2
10 *Cyclostomum tetracanthum* (*Strongylus tetracanthus*)	6 (tetrads)	12 (monads)	undetermined	12	?	1
11 *Cystidicola farionis* (*Ancyracanthus cystidicola*)	6 (tetrads)	11-12 (monads)	undetermined	11-12	X = 1	1
12 *Dictyocaulus filaria* (*Strongylus filaria*)	6 (tetrads)	11-12 (monads)	undetermined	11-12	X = 1*	1
13 *Dictyocaulus viviparus* (*Strongylus micrurus*)	6 (tetrads)	11-12 (monads)	undetermined	11-12	X = 1	1
14 *Diapharynx spiralis* (*Acuaria spiralis*)	6 (di-tetrads)	11-12 (dyads)	undetermined	22-24	X = 1	2
15 *Filaroides mustelarum*	8 (tetrads)	16 (monads)	undetermined	16	?	1
16 *Heterakis dispar*	5 (tetrads)	9-10 (monads)	undetermined	9-10	X = 1	1
17 *Heterakis gallinae* (*Heterakis vesicularis*)	5 (tetrads)	9-10 (monads)	undetermined	9-10	X = 1	1
18 *Heterakis papillosa*	5 (di-tetrads)	9-10 (dyads)	undetermined	18-20	X = 1	2
19 *Heterakis spumosa* (*Ganguleterakis spumosa*)	6 (di-tetrads)	10-12 (dyads)	absent	20-24	X = 2	2
20 Heterakid (from pheasant)	5 (tetrads)	9-10 (monads)	undetermined	9-10	X = 1	1
21 *Mastophorus muris* (*Protospirura muris*)	5 (di-tetrads)	9-10 (dyads)	absent	9-10	X = 1	1
22 *Metastrongylus elongatus* (*Strongylus paradoxus*)	6 (di-tetrads)	11-12 (monads)	absent	11-12	X = 1	1
23 *Nematospira turgida*	6 (di-tetrads)	11-12 (dyads)	absent	22-24	X = 1	2
24 *Ophidascaris filaria* (*Ascaris rubicunda*)	?	?	present	?	?	?
25 *Ophiostoma mucronatum*	6 (di-tetrads)	12 (dyads)	undetermined	24	*	2
26 *Parascaris equorum univalens* (*Ascaris megalocephala*)	1 (rod-shaped)	2 (rod-shaped)	present	51-80	X = 1	26 (X = 9)

Table 7 (Continued)

27	*Parascaris equorum bivalens*	2 (rod-shaped)	4 (rod-shaped)	present	96-104	X = 1	22 (X = 8)
28	*Parascaris equorum trivalens***	3 (rod-shaped)	6 (rod-shaped)	present	?	Y = 1	22-26 (hybrid)
29	*Passalurus ambiguus* (*Oxyuris ambigua*)	4 (tetrads)	7 8 (monads)	undetermined	7-8	X = 1	1
30	*Physaloptera turgida*	5 (di-tetrads)	9-10 (dyads)	absent	18-20	X = 1	2
31	*Proleptus robustus* (*Cornilla robusta*)	8 (tetrads)	16 (monads)	undetermined	16	?	1
32	*Rhabdias bufonis* (*Rhabditis nigrovenosa*)	6 (tetrads) (di-tetrads?)	11-12 (monads) (dyads?)	absent	22 24	X = 1*	2
33	*Rhabdias fulleborni*	6 (tetrads) (di-tetrads?)	11-12 (monads) (dyads?)	absent (?)	22 24	X = 1	2
34	*Rhabditis aberrans* (female - parthenogenetic) (male - non-funtional) (var of *R. aspera*?)	18 (dyads) 9 (tetrads)	18 (monads) 17-18 (monads)	undetermined	18 18	X = 1 X = 1*	1 1
35	*Rhabditis aspera*	7 (tetrads)	13-14 (monads)	undetermined	13-14	X = 1	1
36	*Rhabditis pellio* Butschli (*R. maupasi*)	7 (tetrads)	13-14 (monads)	undetermined	13-14	X = 1	1
37	*Rhabditis pellio* Schneider	7 (tetrads)	13-14 (monads)	undetermined	13-14	X = 1	1
38	*Rhabditis pellio* Schneider (mutant parthenogenetic) (female)	14 (dyads)	11 (monads)	undetermined	14	X = 1	1
39	*Setaria equina* (*Filaria papillosa*)	6 (tetrads)	11-12 (monads)	undetermined	11-12	X = 1	1
40	*Spirina parasitifera*	7 (compound)	14 (compound)	absent	14 (comp)	No "X"	Undetermined, but many
41	*Spirura talpae* (*Spiroptera strumosa*)	8 (tetrads)	16 (monads)	undetermined	16	?	1
42	*Strongylus edentatus* (*Sclerostomum edentatum*)	6 (tetrads)	11-12 (monads)	absent	11 12	X = 1	1
43	*Strongylus equinus* (*Sclerostomum equinum*)	6 (tetrads)	11-12 (monads)	absent	11-12	X = 1	1
44	*Strongylus vulgaris* (*Sclerostomum vulgare*)	6 (tetrads)	11-12 (monads)	absent	11-12	X = 1	1
45	*Syphacia obvelata* (*Oxyurus obvelata*)	8 (tetrads)	15-16 (monads)	absent	15-16	X = 1	1
46	*Toxocara canis* (*Toxascaris canis*)	18 (d tetrads)	30 36 (dyads)	present	60-72	X = 6	2
47	*Toxocara cati* (*Belascaris mystax*)	9 (tetrads)	18 (monads)	present	18	X = 1*	1
48	*Toxocara vulpis* (*Belascaris triquetra*)	12 (di-tetrads)	22-24 (dyads)	present	44-46	X = 2?	2
49	*Trichosomoides crassicauda*	4 (di-tetrads)	7- 8 (dyads)	absent	7- 8 (dyads)	X = 1	2
50	*Trichostrongylus tenuis* (*Strongylus tenuis*)	6 (tetrads)	11-12 (monads)	undetermined	11-12	X = 1	1

* A "Y" chromosome has been reported from these species, but the accuracy of the interpretations is questionable

** Li (1934, 1937) reports a six-chromosome and a nine-chromosome variety of *P. equorum* as occurring in Chinese Mongolian horses. He suggests that these either are examples of polyploidy or represent the primitive racial strains of the species that are still found in the most primitive of living horses. Cytological study of the six-chromosome form shows a behavior normal for *P. equorum* except that the autosomal number present after "diminution" is less than in those found in the better known varieties.

Bibliography

ACKERT, J F 1931 —The morphology nd life history of the fowl nematode, *Ascaridia lineata* (Schneider) Parasitol , v 23 (3) 360-379, 25 figs , pls 13-14

AUERBACH, L 1874 —Organologische Studien Heft 1 2 Zur Charactenstik und Lebensgeschichte der Zellkerne 262 pp , pls 1-4 Breslau

BELAR, K 1923 —Ueber den Chromosomenzyklus von parthenogenetischen Erdnematoden Biol Zentralbl , v 43 (5) 513-518, d figs

BENEDEN, E VAN, 1883 —Recherches sur la maturation de l'œuf, la fecondation et la division cellulaire 424 pp , pls 3, 10-19 bis Gard and Leipzig, Paris

BENEDEN, E VAN, and NEYT, A 1887 —Nouvelles recherches sur la fécondation et la division mitosique chez l'ascaride megalocephale Bull Acad Roy Sc Belgique, 3 Ser , v 14 (8) 215-295, 6 pls

BONNIE, K 1901 — Ueber Chromatindiminution bei Nematoden Jena Zeitschr , 36 n F , v 29 (1-2) 275-285 2 pls figs 1-21
1908 —Chromosomenstudien Arch Zellforsch , v 1 (2-3) 450-514, 2 figs , pls 11-15, 99 figs
1913 —Ueber die Struktur und Genese der Ascarischromosomen Arch Zellforsch , v 9 (3) 433-457, 7 figs

BOVERI, Th 1887 —Zellen-Studien Heft 1 Die Bildung der Richtungskorper bei *Ascaris megalocephala* und *Ascaris lumbricoides* 93 pn 4 pls Jena
1888 —Zellen-Studien Heft 2 Die Befruchtung und Teilung des Eies von *Ascaris megalocephala* 198 pp 5 pls figs 1-94 Jena
1890 —Zellen-Studien Heft 3 Ueber das Verhalten der chromatischen Kernsubstanz bei der Bildung der Richtungskorper und bei der Befruchtung 88 pp 3 pls Jena
1900 —Zellen-Studien Heft 4 Ueber die Natur der Centrosomen 220 pp , figs A-C, 8 pls figs 1-111 Jena
1909 —Die Blastomerenkerne von *Ascaris megalocephala* und die Theorie der Chromosomenindividualitat Arch Zellforsch , v 3 (1-2) 181-268, figs 1-7, pls 7-11, figs 1-51
1911 —Ueber das Verhalten der Geschlechtschromosomen bei Hermaphroditismus Beobachtungen an *Rhabditis nigrovenosa* Verhandl Phys-Med Gesellsch , Würzburg, n F , v 41 (5) 83-97, 19 figs

BOWEN R H 1925 —Further notes on the acrosome of animal sperm The homologies of non-flagellate sperms Anat Rec , v 31 201-231, 5 figs

BRAUER A 1893 —Zur Kenntniss der Spermatogenese von *Ascaris megalocephala* Arch Mikr Anat , v 42 (1) 153 213 3 pls , figs 1-228

CHITWOOD, B G 1931 —Flagellate spermatozoa in a nematode (*Trilobus longus*) J Wash Acad Sci , v 21 (3) 41-42 2 figs

COBB, N A 1925 —Nemic spermatogenesis J Heredity, v 16 (10) 357-359 1 fig
1928 —Nemic spermatogenesis with a suggested discussion of simple organisms -htobionts J Wash Acad Sci v 18 (2) 37-50, 17 figs

DREYFUS A 1937 —Contribuiçao para o estudo do cyclo chromosomico e da determinacao do sexo de *Rhabdias fulleborni* Trav 192b Bol Fac Philos Sci Let, Univ Sao Paulo III Biol Geral No 1, 145 pp , 92 figs

EDWARDS C L 1910 —The idiochromosomes in *Ascaris megalocephala* and *Ascaris lumbricoides* Arch Zellforsch , v 5 (3) 422-429, pls 21-22, figs 1-39

FOGG, L C 1930 —A stud in chromatin diminution in *Ascaris* and *Ephestia* J Morph , v 50 (2) 413-451, 4 pls figs 1-49
1931 —A review of the history of the centriole in the second cleavage of *Ascaris megalocephala bivalens* Anat Rec , v 49 (3) 251-264, 2 pls , figs 1-13

FURST, E 1898 —Ueber Centrosomen bei *Ascaris megalocephala* Arch Mikr Anat , v 52 (1) 97-133 pls 8-9, figs 1 35

GOODRICH H B 1914 —The maturation divisions in *Ascaris incurva* Biol Bull , v 27 (3) 147-150, 1 pl , figs 1-13

1916 —The germ cells in *Ascaris incurva* J Exp Zool , v 21 (1) 61-99, figs, a-k, 3 pls , figs 1-49

GOUILLIART, M 1932 —Le comportement de l'heterochromosome dans la spermatogenese et dans l'ovogenese chez un *Ascaris megalocephala* hermaphrodite Compt Rend Soc Biol , Paris v 110 (28) 117b-1179, 9 figs

GRIGGS, R F 1906 —A reducing division in *Ascaris* Ohio Naturalist v 6 (7) 519-527, pl 33, figs 1-12

HELD, H 1912 —Ueber den Vorgang der Befruchtung bei *Ascaris megalocephala* (In Verhandl Anat Gesellecsh 26 Versamml) Anat Anz , v 41 242-248
1917 —Untersuchungen uber den Vorgang der Befruchtung I Der Anteil des Protoplasmas an der Befruchtung von *Ascaris megalocephala* Arch Mikr Anat , v 89 (2) 59 224 6 pls

HERLA, V 1894 —Etude des variations de la mitose chez l ascaride mégalocephale Arch Biol , Gand v 13 (3) 423-520, pls 15-19, figs 1 103

HERTWIG O 1890 —Vergleich der Ei- und Samenbildung bei Nematoden Arch Mikr Anat , v 36 (1) 1-38, 4 pls , figs 1-18

HERTWIG P 1920 —Abweichende Form der Parthogenese bei einer Mutation von *Rhabditis pellio* Arch Mikr Anat , v 94 303 337, pl 21, figs 1-14

HIRSCHLER, J 1913 —Ueber die Plasmastrukturen (Mitochondrien, Golgischer Apparat, u a) in den Geschlechtszellen der Ascariden Arch Zellforsch , v 9 (3) 351-308, pls 20 21, figs 1 41

JEFFREY, E C and HAERTL, E J 1938 —The nature of certain so-called sex chromosomes in *Ascaris* La Cellule v 47 (2) 237 244, 2 text figs, 1 pl

KEMNITZ, G A 1913 —Eibildung Eireifung, Samenreifung und Befruchtung von *Brachycoelium salamandrae* (*Brachycoelium crassicolle* [Rud]) Arch Zellforsch , v 10 (14) 470-506, pl 39, 37 figs

KRUGER F 1913 —Fortpflanzung und Keimzellenbildung von *Rhabditis aberrans* nov sp Zeit Wiss Zool , v 105 (1) 87-124, pls 3-6, figs 1-73

LI, J C 1934 —A six-chromosome *Ascaris* found in Chinese horses Peking Nat Hist Bull v 9 (2) 131-132 5 figs
1937 —Studies of the chromosomes of *Ascaris megalocephala trivalens* I The occurrence and possible origin of nine-chromosome forms Peking Nat Hist Bull , v 11 (4) 373-379, 1 pl

MAN J G DE 1886 —Anatomische Untersuchungen über freilebende Nordsee Nematoden 82 pp , 13 pls Leipzig
1884a —Espèces et genres nouveaux de nematodes libres de la mer du nord et de la manche Mem Soc Zool France v 2 (1) 1-10
1889b —Troisieme note sur les nematodes libres de la mer du nord et de la manche Mem Soc Zool France, v 2 (3) 182-216, pls 5-9, figs 1-14

MARCUS H 1906a —Ueber die Beweglichkeit der Ascaris-Spermien Biol Centralbl , v 26 (13-15) 427-430, figs 1-5
1906b —Ei- und Samenreife bei *Ascaris canis* Werner (*Ascaris mystax*) Arch Mikr Anat , v 68 (3) 441-490 10 figs , pls 29-30 figs 1-57

MAUPAS, E 1900 —Modes et formes de reproduction des nematodes Arch Zool Exp et Gen , Ser 3, v 8 463-624, pls 16-26

MAYER, A 1908 —Zur Kenntnis der Samenbildung bei *Ascaris megalocephala* Zool Jahrb , Abt Anat , v 25 (3) 495-546, 2 figs , pls 15-16, 61 figs

MEVES, F 1903 —Ueber oligopyrene und apyrene Spermien und ihre Entstehung nach Beobachtungen an *Paludina* und *Pygaera* Arch Mikr Anat , v 61 1-85, 3 pls
1911 —Ueber die Beteiligung der Plastochondrien an der Befruchtung des Eies von *Ascaris megalocephala* Arch Mikr Anat v 76 (4) 683-713 pls 27-29 figs 1-18
1915 —Ueber Mitwirkung der Plastosomen bei der Befruchtung des Eies von *Filaria papillosa* Arch Mikr Anat v 87 (2) 12-46, 4 pls , figs 1-77
1920 —Ueber Samenbildung und Befruchtung bei *Oryuris ambigua* Arch Mikr Anat , v 94 135-184, pls 9-13, 72 figs

214

MEYER, O 1895 —Cellulare Untersuchungen an Nematoden Eiern Jena Zeitschr, v 29 (N F, v 22) 391-410 2 pls

MULSOW, K 1911 —Chromosomenverhaltnisse bei Ancyracanthus cystidicola Zool Anz, v 38 (22-23) 484-486, 6 figs
1912 —Der Chromosomencyclus bei Ancyracanthus cystidicola (Rud) Arch Zellforsch, v 9 (1) 63-72, 5 figs, pls 5 6, figs 1-45

POTTS, F A 1910 —Notes on the free-living nematodes Quart J Micr Sc, (n s) v 55 (219) 433 484, 11 figs

PABL, C 1889 —Ueber Zellteilung Anat Anz, v 4 (1) 21-30, 2 figs

ROMEIS, B 1912 —Beobachtungen uber Degenerationserscheinungen von Chondriosomen Nach untersuchungen an nicht zur Befruchtung gelangten Spermien von Ascaris megalocephala Arch Mikr Anat, v 80 (Abt 2) (4) 129-170, 2 pls

ROMIEU, M 1911 — Le Spermiogenese chez l'Ascaris megalocephala Arch Zellforsch, v 6 (2) 254-325, pls 14-17, figs 1-94

SALA L 1895 —Experimentelle Untersuchungen uber die Reifung und Befruchtung der Eier bei Ascaris megalocephala Arch Mikr Anat, v 44 (3) 422-498, pls 25-29, figs 1-89

SCHEBEN, I 1905 —Beitrage zur Kenntnis des Spermatozoons von Ascaris megalocephala Ztschr Wiss Zool, v 79 (3) 397 431 3 figs pls 20-21, figs 1-44

SCHLEIP, W 1911 —Das Verhalten des Chromatins bei Angiostomum (Rhabdonema) nigrovenosum Ein Beitrag zur Kenntnis der Beziehungen zwischen Chromatin und Geschlechtsbestimmung Arch Zellforsch, v 7 (1) 87-138, pls 4-8, figs 1-108
1924 —Der Herkunft der Polaritat des Eies von Ascaris megalocephala Arch Mikr Anat Entw-Mech, v 100 (3) 573-598, 17 figs

SPEK, J 1918a — Oberflachenspannungs-differenzen als Ursache der Zellteilung Arch Entw-Mech, v 44 (1) 1-113 25 figs
1918b —Die amoboiden Bewegungen und Stromungen in den Eizellen einiger Nematoden wahrend der Vereinigung der Vorkerne Arch Entw-Mech, v 44 (2) 217-255, 15 figs

1920 —Experimentalle Beitrage zur Kolloidchemie der Zellteilung Kolloidchem Beih, v 12 (1-3) 1-91

STRUCKMANN, C 1905 —Eibildung, Samenbildung und Befruchtung von Strongylus filaria Zool Jahrb, Abt Anat, v 22 (3) 577-628, pls 29-31, figs 1-105

STURDIVANT H P 1931 —Central bodies in the spermforming divisions of Ascaris Science, n s v 73 (1894) 417-418
1934 —Studies on the spermatocyte divisions in Ascaris megalocephala, with special reference to the central bodies, Golgi complex and Mitochondria J Morph, v 55 (3) 485-475, pls 1-5, 81 figs

TRETIAKOV, D 1904 —Die Spermatogenese bei Ascaris megalocephala Arch Mikr Anat, v 65 (2) 383-438, 1 fig, pls 22-24, figs 1-130

VEJDOVSKY F, 1911-1912 —Zum Problem der Vererbungstrager 184 pp, 16 figs, 12 pls Prag

WALTON A C 1916a —Ascaris canis (Werner) and Ascaris felis (Goeze) A taxonomic and a cytological comparison Biol Bull, v 31 (5) 364-372, figs A-F, 1 pl, figs 1-12
1916b —The "refractive body" and the "mitochondria" of Ascaris canis Werner Proc Amer Acad Arts and Sci, v 52 (5) 253 266, 2 pls, figs 1-16
1918 —The oogenesis and early embryology of Ascaris canis Werner J Morph, v 30 (2) 527-603, 1 fig, 9 pls, figs 1-81
1924 —Studies on Nematode Gametogenesis Ztschr Zell-u Geweb v 1 (2) 167-239, figs A-B, pls 8-11, figs 1-118

WARREN, E 1930 —Multiple spermatozoa and the chromosome hypothesis of heredity Nature (Lond), v 126 (3166) 973-974, figs 1-9

WILDMAN, E E 1913 —The spermatogenesis of Ascaris megalocephala, with special reference to the two cytoplasmic inclusions, the refractive body and the "mitochondria", their origin nature and rôle in fertilization J Morph, v 24 (3) 421-457, 3 pls, 48 figs

WILSON, E B 1925 —The cell in development and heredity 3d ed xxxvii & 1232 pp, 529 figs New York

ZOJA, R 1896 —Untersuchungen uber die Entwicklung der Ascaris megalocephala Arch Mikr Anat, v 47 (2) 218 260, pls 13-14

CHAPTER II

NEMIC EMBRYOLOGY

B G CHITWOOD

Nemic embryology is a subject which has stimulated much research, especially because of the fact that the cells designed to form particular organs are laid down in the very early cleavages This type of development is termed determinate cleavage and in substance means that each blastomere may be identified in the egg as the stem cell of a particular organ or part of an organ In other words, the fate of each cell is foreordained from the first division

The regularity with which division takes place in nematodes was observed by the earliest workers on the subject No attempt will be made to give an historical account of the development of our knowledge other than to point out a few of the steps Butschli (1875), Galeb (1878), Goette (1882), and Hallez (1885) were among the pioneers in the field and to them the later workers are indebted for breaking the ice, but in the light of present day knowledge their observations appear rather casual The publication of Boveri in 1892 on the embryology of Parascaris equorum was the foundation of modern nemic embryology His investigations were followed by those of zur Strassen (1892, 1896), Spemann (1895), Zoja (1896), Neuhaus (1903), Mueller (1903), Martini (1903, 1909), and Pai (1927, 1928) as well as many less comprehensive studies by other authors It should be stated that Boveri's work directly initiated the present study of the subject by later workers, all these investigations have given us information equalled in few other groups of animals

Nemic embryology consists of the study of individual cells, in the early cleavages each cell is differentiated to such an extent that it is capable of giving rise only to certain parts of the organism, sister cells as a rule differ to some degree in their potentialities While there is some difference in opinion as to what some particular cells may give rise to in the mature organism, these differences appear to be based more upon conceptions of authors than actual conditions in given species It is not surprising that misinterpretations should arise in the study of cell lineage where one must follow the course of literally hundreds of cells

In nemas the development of germ layers as observed in other animals is highly modified In fact one can hardly speak of germ layers in reference to nemas In the course of the first cleavages a number of so called "primordial" or "stem" cells are formed (Fig 151) These are highly differentiated as to their potentialities Each of them will form a certain organ or organ system, e g, the anterior cell is destined to form the greater part of the ectodermal epithelium and is designated S1, which means first "somatic" stem cell, this cell therefore is the ancestor of the primary ectoderm The other cell, posterior in position, however, is a less differentiated cell in its potentiality It forms the remainder of the embryo, for this reason it is designated P1, or first parental germinal cell, the fertilized ovum being designated P0 This first cleavage is a transverse one These "primordial" or "stem" cells are therefore unequal in their prospective potencies

Beginning with the second cleavage the cells of each given family have their own cleavage "rhythm", that is, cells descendent from each primordial or stem cell divide at the same rate but they differ in the rate from those descended from another primordial or stem cell Ordinarily one would expect this to be due to a difference in the amount of yolk or the size of the cells but in nematodes this is not the case, instead the cell has an inherent rate of cleavage without regard to size or yolk and it cannot be explained as caused by mechanical forces

The second cleavage is transverse in the S1 cell forming an anterior dorsal cell A and a posterior dorsal cell B This is followed by transverse division of the blastomere P1 forming an anterior ventral cell S2 and a posterior cell P2 The second somatic stem (S2) cell is destined to

form the somatic musculature, part of the esophagus, and the entire intestine or mesenteron

At the third cleavage the S1 cell group, A and B, divides longitudinally forming four cells—two on the right and two on the left side of the embryo, the cells on the right are designated by Roman letters a and b, while those on the left are designated by Greek letters Alpha and Beta, for which the small capitals A and B are substituted in the text Following this the second somatic stem cell, S2, divides transversely, the anterior cell being destined to form the greater part of the mesoderm or the body wall and esophagus, termed MST, and the posterior cell the entire entoderm, designated E The undifferentiated stem cell P2 divides transversely, the dorsal posterior daughter being termed S3 or the third somatic stem cell, and the posterior ventral cell P3 The third somatic stem cell is destined to form the ectodermal epithelium of the posterior part of the body and is therefore secondary ectoderm it also forms a part of the mesoderm in some nemas

The next or fourth cleavage is commonly said to complete the formation of stem cells going into the formation of the soma or body The cleavage of the S1 group is transverse forming a I, a II, a I, a II, b I, b II, and b II, of the S2 group M divides obliquely, transversely or longitudinally forming either two cells one behind the other St and AI or two cells side by side mst and MST, E divides transversely forming E I and E II, the S3 cell group divides longitudinally forming e and c, and finally the P3 cell divides transversely forming P4, ventral, and S4, posterior The descendants of S4 are destined to form the proctodeum or rectum and sometimes the mesoderm or ectoderm of the posterior entral part and therefore either tertiary ectoderm, or possibly secondary mesoderm (see description of embryology of Parascaris equorum)

In further cleavages the descendants of given cells are designated by Roman and Greek letters (small capitals in the text) if they go to opposite sides of the embryo, i e, a and a, by Roman numerals if serial, i e, I and II, and by Arabic numerals if neither, i e, I and 2 Thus a longitudinal division of S4 (D) forms cells d and D, transverse division of them forms d I, d II, D I and D II Additional labels such as a', a'' etc, are sometimes necessary P4 is generally termed the primordial germ cell but recent investigations indicate this may not be the case The epithelium of the reproductive system is not germinal tissue but somatic tissue and is probably laid down by a later cleavage At the fifth cleavage two cells are formed which are variously termed G I and G II and S5 and P5, these cells appear alike The earliest differentiated genital primordium contains four cells,—two epithelial and two germinal (G I and G II) The former may be of mesenchymatous origin or they may have been derived from the P line at either the fifth or sixth cleavage

The various cleavages as outlined above do not take place in all cells simultaneously so we do not have a regular doubling of cells Though the cleavage is total or holoblastic, it is usually unequal, since it is neither radial nor spiral there is no typical morula stage There may or may not be a segmentation cavity or blastocoele and when such is present it is usually of small size As Martini (1908) showed, the absence of a segmentation cavity is a negative point being in these forms entirely dependent upon the depth of the cleavage furrows, the depth of furrows does not appear to be correlated with anything else in the cleavage of nemic ova and embryos so that it must be regarded as without significance Generally speaking we may say that the blastular stage begins in the 12 to 16 cell stages of the embryo since it is at this time that a blastocoele appears in certain species, Parascaris equorum, Rhabdias bufonis, and Nematoxys ornatus, while embryos composed of homologous cells in the species Camallanus lacustris and

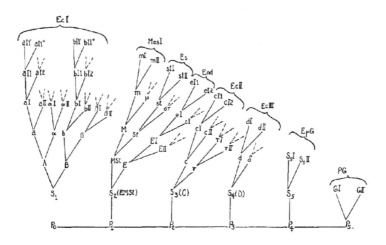

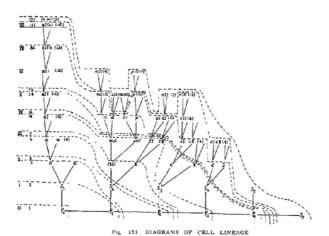

Fig. 151 DIAGRAMS OF CELL LINEAGE

Upper figure is a general diagram showing the standard numbering system of neural blastomeres
Lower figure is a diagram of cleavage in *Rhabditis bufonis*. Roman numerals in the first vertical column indicate number of cleavage. Arabic numerals in first vertical column give total number of cells in embryo. Numbers in parenthesis indicate number of cells in a given germ line. Broken curves indicate corresponding levels of cleavage in the various germinal lines. Original

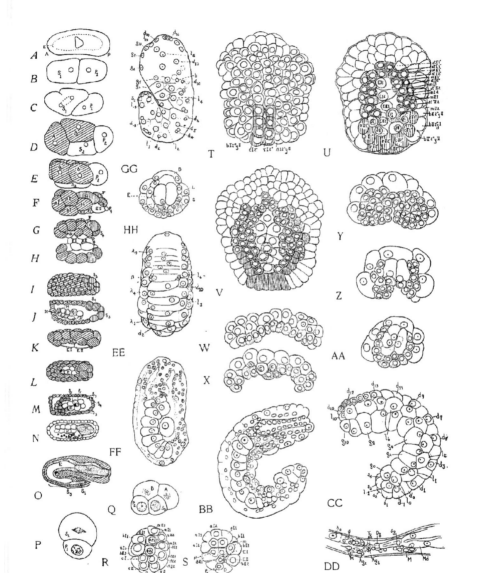

Fig. 152.

Pseudalius minos do not have a blastocoele Two layered cell plates such as occur in the latter instance were at one time considered a type of gastrula, sterrogastrula, but since this stage is followed by epiboly characteristic of gastrulation it must be considered a type of blastula for which the term *placula* has been used

Gastrulation (Figs 153 L-Q), on the other hand, is the entrance of the entoderm and mesoderm into a hull surrounded by ectoderm, this being completed at the closure of the blastopore Dependent upon the presence and size of the absence of a blastocoele there are two possible ways in which the ectoderm may come to surround entoderm (Martini, 1908), (1) the cells may retain their relative positions (synectic) or they may not retain their relative positions (apolytic) In the absence of a blastocoele the ectoderm may grow over the entoderm either with or without change in cell positions so we may have epibolic synectic or epibolic apolytic gastrulation Invagination, that is embolic gastrulation is possible only if there is a blastocoele and in this case it may be either apolytic or synectic Epibolic apolytic gastrulation is not known among nematodes but the other possibilities are represented *Parascaris equorum* undergoes embolic apolytic gastrulation, *Rhabdias* and *Nematoxys* embolic synectic, and *Camallanus* and *Pseudalius* epibolic synectic (Fig 153 P)

Regarding the development of the mesoderm in nematodes there are certain points of interest The mesodermal stem cells at the time of gastrulation are arranged in rows on either side of the entoderm as well as anterior to it They follow the entoderm in sinking into the primary body cavity and later form two sub dorsal and two subventral strings on either side of the entoderm Since the individual cells maintain their identity and no cavity is formed between them it would not be proper to say that nematodes had a true coelome The individual cells would properly be termed a mesenchyme However, certain mesodermal cells do cover the organs in the body cavity and for that reason we may say the body cavity is analogous but not homologous to a coelome, i e, a pseudocoelome The mesoderm may be said to have been derived from the entoderm since it comes chiefly from a cell (S2) which also forms the entoderm but in *Parascaris equorum* as well as some other nematodes part of the derivatives of the mesodermal stem cell (MSt) enter into the formation of the ectoderm according to some authors Strictly speaking it would seem preferable to consider six germ layers the ectoderm, mesoderm (somatic musculature and isolation tissue), entoderm, esophagus (St-S1), somatic part of gonad (S5) and germinal layer (P5)

The S1 cell group in nematodes forms the greater part of the epithelium which is usually so arranged that the nuclei and the cell bodies of the cells are situated in the dorsal ventral, and lateral chords It also contributes to the formation of the esophagus, the nervous system and the excretory system

The S2 cell group forms the greater part of the mesoderm, that is the longitudinal muscles, transverse muscles, and probably the isolation tissue, as well as the musculature of the esophagus It also forms the mesenterion or intestine

The S3 cell group forms the ectodermal epithelium of the posterior part of the body and contributes to the formation of the nervous system and the musculature of that part of the body

The S4 cell group is known to form the rectum and rectal glands and may also form some muscular tissue

The S5 cell group according to Pai (1928) forms the outer covering of the gonads and the epithelium of the

gonoducts If nematodes can be said to have the homologue of a true coelome it would be the lumen of the gonoducts since it is formed as a cavity between cells, but the positional relationship of the S5 group with the entoderm makes its consideration as mesoderm rather questionable

At the time of hatching from the egg nematodes are fully formed, the tissues differentiated and functional with the exception of the reproductive system In some nematodes no further division of cells takes place except in the gonads and structures either directly or indirectly connected with reproduction In all instances known the chords are cellular rather than syncytial, there being five rows of cells in the anterior part of the body of most nematodes studied (a dorsal two lateral, and two ventral) while in the remainder of the body there are two dorsolateral, two lateral, two ventrolateral, and two ventral, the dorsal being absent but there is a thickening of the hypodermis in the dorsal region Changes from this condition take place during later development or not at all

The somatic musculature is platymyarian and meromyarian in the newly hatched larva but it may become coelomyarian and polymyarian later In such an instance we have to recognize the division of functional muscle cells certainly highly differentiated

The esophagus of the larva is similar to that of the adult in some forms, whether or not multiplication of cells may later take place is unknown

The intestine of the larva is composed usually of 2 rows of cells, the lumen being formed by a separation of parallel rows of cells rather than from the archenteron This formation of the lumen seems to be of no general significance since it is the only means by which a lumen could develop in forms with so little blastocoele

The nervous system at least in some forms appears to be of the same number of cells in the larva as in the adult, with the exception of cells innervating genital papillae

The excretory system of the larva is the one system of which our knowledge is entirely inadequate It is usually stated to be formed by a single cell of the ectoderm (S1) but its development has not been satisfactorily traced

Regarding the embryology of particular nematodes, we find that thus far no member of the Aphasmidia has been studied though many members of the Phasmidia have These belong to several diverse groups, *Rhabdias bufonis Rhabditis terricola* (partially), *Diplogaster longicauda* (partially) and *Turbatrix aceti* among the Rhabditina, *Metastrongylus elongatus, Pseudalius inflerus* and *P minor* among the Strongylina, *Ascaris lumbricoides* (partially), *Toxocara canis* (partially) *Parascaris equorum, Nematoxys ornatus,* and *Syphacia obvelata* among the Ascaridina, and *Camallanus lacustris* of the Camallanina Of these forms it appears best to limit our descriptions to *Turbatrix aceti, Rhabdias bufonis, Parascaris equorum,* and *Camallanus lacustris,* comparing other forms with them whenever it appears advisable

Turbatrix aceti (Fig 152) Pai (1927) worked out the development of this species in a very complete manner He found that the end of the ovum at which the sperm entered is destined to form the anterior end of the embryo From the first cleavage which is transverse, two very slightly unequal cells result a larger anterior one designated as S1, and a smaller posterior one, P1 By the second cleavage S1 dividing somewhat horizontally and obliquely gives rise to a dorsal cell B and an anterior cell A Division of P1 follows shortly a ventral cell S2 and a posterior cell P2 resulting The second somatic stem cell S2, is said to form the esophagus, the intestine or mesenterion, and the mesoderm, and for that reason may be designated EMSt i e, entoderm mesoderm, stomodeal stem cell At the third cleavage the S1 cells, A and B divide longitudinally giving the embryo a distinct bilateral symmetry which it retains throughout the remainder of its development This is followed by a transverse division of the ventral cell S2, forming an anterior ventral cell MSt and a posterior ventral cell E Thus in the seven cell stage there are four dorsal ectodermal cells, two ventral ento-mesodermal cells derived from the second somatic stem cell, and a single large posterior undifferentiated cell, P2 In later divisions the blastomeres proceed at an even more unequal rate, there being a distinct tendency for the S2

Fig 152

A O—*Turbtrix aceti* (A—Fertilized ovum B—2 cell stage C—Beginning second cleavage D—4 cell stage F—b cell F—16 cell C H—24 cell I-J—141 cell K—2b cell L—S2 cell M—171 cell N—Early definitive embryo O—[adpole stage] P DD—*Camallanus lacustris* (P—2 cell stage Q—1 cell R—28 cell dorsal view S—28 cell ventral view T—177 cell dorsal view U—177 cell ventral view V—354 cell ventral view W—Cross sections of embryo slightly older than V W—anterior region X—Posterior region Y AA—Cross sections of anterior, mid and posterior regions of still older embryo BB CC—Sagitta and surface views of early definitive embryo DD—Anal region of mature larva EE HH—*Rhabdias bufonis* EE FF—Surface and sagittal views of early definitive embryo GG—Surface view of tadpole stage HH—Cross section of stage shown in FF and FF) A O After Pai 1928 *Zischr Wiss Zool* v 131 (2) ? AA After Martini 1903 Idem v 74 (4) BB DD After Martini 1906 Idem v 81 (4) EE HH After Martini 1907 v 86 (1)

and P series to lag behind the other cells The $S1$ cell group again divides, A and B on the left side, a and b on the right side, each giving rise through an oblique division to two cells, making four on each side of the body At about the same time $P2$ undergoes an unequal division, the larger daughter cell (the third somatic stem cell, $S3$) being dorsal, and the small germ cell (P3) being ventral, the ventral cells MSt and E both divide MSt longitudinally giving rise to MST (left) and mst (right), and E dividing transversely gives rise to E I (anterior) and E II (posterior) The posterior dorsal cell, $S3$ like $S1$ is an ectodermal stem cell It divides longitudinally forming c (left) and c (right), these cells produce the hypoderm s of the posterior part of the embryo The germ cell $P3$ again undergoes unequal transverse division, forming a larger dorsal cell, $S4$, and a smaller ventral cell, $P4$

These cleavages, four in number bring the embryo to the 16 cell stage at which time all of the somatic stem cells have been formed (Fig 152 F) The embryo is an elongate blastula with an inconspicuous blastocoele Cells destined to form the ectoderm cover the anterior and dorsal surfaces of the embryo, cells destined to form the mesoderm and entoderm cover the posterior ventral surfaces In later cleavages it is somewhat easier to follow the fate of the cells of each stem line separately

At the 24 cell stage $S1$ consists of 16 cells (Fig 152 G-H) six on the left side of the embryo, four in the center, and six on the right side The origin of the mediodorsal row of four cells has not been determined Between this stage and the 141 cell stage (Fig 152 I-J) gastrulation is completed, descendants of $S1$ come to cover the anterior two-thirds of the embryo, the cells small and numbering 115 The remainder of the embryo is covered by descendants of $S3$ In further development $S1$ comes to make up four-fifths of the hypodermis, gives rise to the nervous system and in postembryonic development to the vulva (vagina)

Returning to the $S2$ line of somatic cells, the four cells present in the 16 cell stage, MST, mst, E I and E II, proceed at unequal rates At the 26 cell stage there are four cells derived from MSt, namely, ST, st, M, and m, formed by transverse division of the bilaterally symmetrical cells MST and mst By the 32 cell stage the six $S2$ cells form a total of ten, in paired bilateral rows of 5 cells as follows ST, st, M I, m I, M II, m II, E I, e I, E II, e II At this time the entire $S2$ cell group is somewhat sunken inward, making a ventral groove which is completely covered by ectoderm at the 141 cell stage, and gastrulation is completed By the 171 cell stage there are four cells derived from $s1$ and st, eight formed from M, and seven from E Shortly thereafter there are eight entodermal cells The M line (mesoderm) lies in the body cavity By the time the embryo takes a definite vermiform shape (Fig 152 O) there are 12 entodermal cells (E)

At the termination of gastrulation of the 141 cell stage, the third somatic stem line, $S3$, consists of 11 cells The first stage larva, at hatching, consists 15 cells of $S3$ origin which form approximately one-fifth of the hypodermis, for they cover dorsal-posterior, and postanal parts of the body

The fourth line of somatic stem cells $S4$, which originated at the fourth cleavage or 16 cell stage consists of four cells at the termination of gastrulation It forms the tertiary ectoderm (Ec III) which gives rise to the proctodeum or rectum

The germ cell line represented by $P4$ passes a quiescent stage during gastrulation A cleavage takes place shortly before the 171 cell stage forming two cells, one of which ($P5$) produces the reproductive cells, while the other ($S5$) produces the somatic part of the reproductive system Shortly before hatching both $P5$ and $S5$ divide, forming $S5$ I anterior, $S5$ II posterior, G I and G II The $S5$ cells surround the $P5$ cells and are generally termed "terminal" or "cap" cells In the male the entire vas deferens and seminal vesicle are formed later on by $S5$ I, while $S5$ II forms the epithelium of the testis In the female $S5$ I forms the somatic part of the ovary while $S5$ II forms the oviduct, uterus, and seminal receptacle

The entire development from fertilization to formation of the larva within the egg requires two days The larva is "born" three days later At this time the larva

possesses the same number of cells as the adult in all systems except the hypodermis and reproductive system The female is mature on the 6th to 7th, the male on the 9th day after birth, specimens of both sexes may live 49 and 48 days respectively

The gastrulation being somewhat atypical, there is difference of opinion as to the names of germ layers to be applied to the various somatic stem cells Pai regards the St cell group which later forms the esophagus as secondary entoderm but since it is comparable to the M cell group it is better termed mesoderm The $S5$ group, forming the somatic part of the reproductive system, he also terms entoderm, though mesoderm would appear preferable Study beyond the "tadpole" stage (Fig 152 O) is difficult and thus far has been carried out only by means of totomount preparations which lends uncertainty as to the results The following is a catalogue of the cells of the adult

Table 8 Derivation of cells in *Turbatrix aceti*

Stem cell	Structure		Number of cells
$S1$ 4/5 of ectoderm			?
	Nervous system		
	Dorsal anterior to nerve ring		23
	" posterior to nerve ring		9
	" above bulb		5
	" subdorsal cephalic ganglia 2 @ 6		12
	Lateral ganglia	2 @ 28	56
	Ventral subventral ganglia posterior to nerve		
	ring		38
	" retrovesicular ganglion		17
	" anterior to nerve ring		13
	" subventral cephalic ganglia	2 @ 7	14
	" nerve		64
	Excretory cell		1
$S2$	(EMSt)		
	Esophagus		
	Corpus		85
	Isthmus		0
	Bulb		24
	Esophago intestinal valve		?
	Intestine		18
	Musculature		64
	Connective tissue		16
$S3$	(Secondary ectoderm)		
	Hypoderms of dorsal and postanal regions		
		about	5
$S4$	(Tertiary ectoderm)		
	Rectum		20
$S5$	(Only partially determined)		
	Seminal vesicle		24
	Ejaculatory duct		8
	Other structures		?

Rhabdias bufonis The embryology of this species was described by Metschnikoff (1865), Goette (1882), Neuhaus (1903), Ziegler (1895), and Martini (1907) Of these studies those of Metschnikoff were rather casual Goette committed an unfortunate error in incorrectly orienting the early stages of the embryo, the anterior end being considered the posterior and vice-versa

The first cleavage of this species differs from that of *Turbatrix* because it is more nearly transverse due to the difference in shape of the egg Ziegler traced the embryology partially through the eighth cleavage In most respects his results correspond to those obtained by Pai However, there are some differences Martini has given more exact data regarding the late embryology than are known in the case of *Turbatrix aceti*

During the early stages, the embryology of *Rhabdias* is nearly identical with that of *Turbatrix* The following differences have been noted $P2$ divides (third cleavage) before the fourth cleavage begins in the $S1$ group (Fig 151), $S2$ divides longitudinally instead of transversely at the fourth cleavage forming C I and C II At the fifth cleavage the embryo consists of a 30 cell blastula (16 $S1$ cells 2 St cells, 2 M cells, 4 E cells, 4 $S3$, $S4$ and $P4$ cells The fourth somatic stem cell does not divide until after the sixth cleavage has taken place in the $S1$ group forming 32 primary ectodermal cells At this time $S4$ divides longitudinally forming D and d At the seventh cleavage, the $S1$ line totals 64 cells, the M line 8, the St line 4, the E line 4, C line 8, D line 4 and P line ($S5$ plus $P5$ or $P4\ I$ and $P4\ II$) 2 cells, giving a total of 94 cells. Apparently the eighth cleavage is limited to the $S1$ group at this time. Subsequently a ninth cleavage and at least a partial tenth cleavage takes place.

Before becoming elongated the $S1$ group probably is composed of 248 cells. At this time there is a rest from cleavage in the $S1$ group during which the other groups (C, D and E) evidently pass through at least some of the cleavages which they have missed, for the mature larva ready to hatch is composed of between 400 and 500 cells (Martini, 1907).

The posterior extremity of the embryo (Fig 152 EE-HH) begins to bend ventrally and anteriorly at which time the esophagus is well formed, the intestine composed of two rows of seven cells, there are lateral mesodermal chords and a pair of ventral mesodermal chords, the genital primordium is ventral to the intestine and the two cells lie beside one another. The ectodermal cells forming the dorsal and anterior parts of the embryo are larger than the others.

During this period of elongation (Fig 153 D-J & RR) several changes take place. The first two intestinal cells divide longitudinally forming a lumen surrounded by four cells. At this time the genital primordium lies under the sixth and seventh entodermal cells. At hatching the intestine is composed of 20 cells four at the base of the esophagus and two rows of eight cells behind them. The lumen is zigzag but it later becomes wavy. The cells are in two more or less dorsal and ventral rows. At this time according to Martini (1907) the gonad is situated between the twelfth and thirteenth intestinal cells and is composed of about 10 cells, (Fig 153 D).

Further differentiation of the epithelium takes place during the same period. Whereas before elongation the embryo is surrounded by two subdorsal, two dorsolateral, and two lateral to ventrolateral rows of large cells and numerous small ventral cells, a distinct rearrangement now takes place. The subdorsal rows which are at first opposite come to be alternate (Fig 152 EE). (It should be noted that subsequent lettering of cells has no correlation with the lettering used to refer to cells during the first seven cleavages.) There is a gradual pushing of the subventral rows squeezing some of the ventral small cells into the body cavity. Slightly later the embryo takes a vermiform appearance commonly called the "tadpole stage" (Fig 152 GG). By this time the dorsal row of cells is split, it extends from what is now the swollen region corresponding roughly to the position of the nerve ring, to slightly anterior to the level of the anus. The embryo is left without nuclei in the dorsal line in this whole region. The lateral mesodermal chords (Fig 153 G) become dorsosubmedian and form the submedian muscle fields. Some of the cells previously ventral form the subventral muscle fields. These mesodermal tissues pressing against the epithelial cells in the submedian areas cause the six rows of previously mentioned large ectodermal cells to be pressed laterally forming the lateral chords. In the stage shown in Fig 152 GG there is an anterior mediodorsal row of seven cells (derived from $S3$ group) which do not separate (d 14-20) but remain as the nuclei of the dorsal chord, the next posterior-most cell, $d\ 13$ goes to the right, $d\ 12$ to the left and so on. They form the dorsolateral parts of the lateral chords. Two small epithelial cells b and B are covered by $d\ 10$. Posterior to the twenty-second dorsal cell, d-1 and $g\ 0$, $G\ 0$, l-1 and L-1 there are four unpaired cells. The cells destined to form the lateral cell rows of the lateral chords ($l\ 1$, 2 etc of Fig 152 G(r)) number eleven on each side, $l\ 7$-10 being in the cephalic region, $l\ 1$-8 being in the mid region, and l-1, postanal. The cells destined to form the ventrolateral part of the lateral chords ($g\ 0$-10) also number eleven, $g\ 8$-10 forming subventral rows in the cephalic region while the other cells are already in final position. The anus is posterior to $g\ 1$ and $G\ 1$. The greater part of the nerve cells and the cells forming the ventral chord come from the small cells on the ventral surface of the embryo.

At hatching the dorsal chord has a single row of nuclei confined to the cephalic region, in which region the lateral chords also have a single row of nuclei, the ventral, two rows. Posterior to the cephalic region the lateral chords have three rows of nuclei each, the ventral a questionable number.

Camallanus lacustris (Fig 152 & 153). The embryology of this form described under the name *Cucullanus elegans* has been studied by Bütschli (1875) and Martini (1903, 1906). There are several minor variations in the form of early cleavage from that seen in the previously studied forms and the development has been followed somewhat further.

The first cleavage forming $S1$ and $P1$ is very unequal (Fig 152 P). The four-cell stage is rhomboid at its formation as also in *Rhabdias*. All of the following cleavages are characterized by smaller furrows than in previous forms and no blastocoele is developed. Subsequent cleavages (Fig 152 Q-S) are similar to those in *Turbatrix aceti* and will be omitted up to the initiation of the ninth cleavage.

The ninth cleavage of the $S1$ group forms 256 cells, ninth of the C group 32 cells, ninth of the St group 32, eighth M 16 cells, seventh of D 8 cells which, together with the 8 cells of the E group and 2 of the $P4$ group, forms an embryo of 354 cells. This represents a gastrula (Fig 152 V) the rim of which is formed by several rows made up of the St and M groups anteriad, and by the D and C groups lateriad and posteriad.

Following the 354-cell stage the St cell group divides (10th cleavage) forming 64 cells, the M groups divide forming 32 cells (9th cleavage). The F group divides forming 16 cells (7th cleavage), the C cell groups divide in part forming about 48 cells (9th cleavage), the D cell groups divide to form 16 cells (8th cleavage), and the primordial germ cell $P4$ divides. It is said that a part of the $S1$ cell group may also divide but this is uncertain. The resulting embryo consists of approximately 486 cells.

Gastrulation occurs between the 354 and 486 cell stages. The dorsal surface is convex, the ventral surface concave, in the anterior part of the embryo the curve is most pronounced in the median line (Fig 152 W) while in the posterior part it is most marked toward the edges (Fig 152 X). This becomes more outspoken with age and is correlated with swelling of the dorsal ectodermal cell rows. At this time the dorsal surface of the embryo is covered by 6 longitudinal rows of large cells, derivatives of $S1$ and C and the sides, anterior and posterior ends are covered by smaller cells derived from the same cell groups (Fig 152 Y-AA). Gradually these 6 dorsal cell rows come to cover the smaller ventral cell rows which themselves cover the M, St and E cell groups. The gastrulation is thus through epiboly. The St and M cell groups are pushed in the groove becoming closed at the anterior end, and there are some cells of the $S1$ group which enter the inside of the embryo at the anterior end. At the stage represented in figure 152 Z, the posterior part of the ventral groove is open. Finally closure of the posterior part of the ventral groove takes place, the large dorsal cell rows coming to surround the small ventral cells of the D cell group and some of the C cell group (Fig 152 CC). At the same time the two most dorsal cell rows fuse so that the embryo is covered by 6 cell rows.

Organogenesis. It has not been possible to follow the history of individual cells during their rearrangement at the completion of gastrulation. Because of this, a new nomenclature is adopted to mark the shapes of further development. The embryo apparently does not increase in number of cells but the cells become differentiated into organs. The embryo which is elongate or sausage shaped is covered by a mediodorsal cell row ($d\ 1$-20 and $s\ 1$-4), 2 lateral rows (l and L -1-10), and 2 subventral rows (g and G 0-10). All of these large cell rows are probably derived from C. The "0" designates the position or level of the future proctodeum (Fig 152 CC). The ventral and anterior small cells of the $S1$ cell group, the ventral small cells of the $S3$ cell group, the M, St and M cell groups as well as the descendents of $P4$ are all enclosed by the epithelium formed by the above mentioned cell rows. The anterior end of the embryo is covered by five cells, $d\ 20\ i\ 10\ L\ 10$, $g\ 10$, and $l\ 10$. The primordia of all organs are definitely recognizable.

Formation of the dorsal, ventral, and lateral chords takes place in the following manner. The mesodermal strands push against the outer cell layers in the four

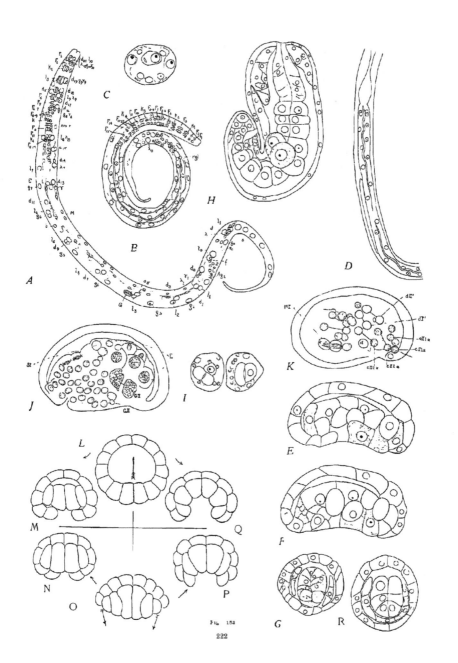

Fig. 152

submedian regions In the anterior part of the body this causes the l and r cell groups to be incompletely separated from the d cell row dorsally and the g and c cell rows ventrally The result is that there are four chords, the dorsal and two lateral chords consisting of one cell row each, and the ventral chord of two cell rows This takes place in the section of the embryo covered by d 1-10, l and r 8-10 g and c 8-10 In the remainder of the embryo the primordia of the muscles press and separate the alternate cells of the dorsal cell row and the rows of the two ventral cell rows This causes the formation of two lateral rows of three cells each the dorso lateral cell rows being formed by d and p cells, the ventrolateral by g and c cells, the result being two lateral chords of three cell rows There remains a thickening of the mediodorsal part of the epidermis which is free of nuclei, the dorsal chord and a ventral thickening which contains the small Sl and Sr cells forming the ventral chord with its ganglia

The mesoderm giving rise to the subdorsal and subventral muscle bands is derived chiefly from cells of the M group but posterior cells of the St group also contribute As they push out between the covering epithelial cells they become completely differentiated, and form overlapping double rows of platymyarian muscles in each sector

Immediately after the closure of the ventral groove we find a neatly solid mass of cells anterior to the intestine This is the primordium of the esophagus (Fig 152 BB) It has apparently arisen from two cell groups the St and small cells of the Sl group The lumen of the esophagus has in its origin no connection whatever with the ventral groove The cells enter the body cavity as a mass becoming arranged in a triradiate pattern The lumen is formed by separation of the cells Already in the stage represented above, the various cells may be recognized which are later present in the adult The nuclei are very closely placed behind one another, there being a total of 66

As the embryo becomes more elongated the nuclei are separated and at hatching come to form an esophagus consisting of an anterior part containing two groups of three marginal nuclei, two groups of six radial nuclei (Fig 153 A) and a posterior part containing two groups or three marginal nuclei two groups of six radial nuclei two groups of two subventral radial nuclei, six groups of three radial nuclei (Fig 153 A) The same number of nuclei was observed in the adult stage Part of the radial nuclei are probably nuclei of the esophageal glands and part nuclei of the esophago-sympathetic nervous system Martini considers that the gland cells marginal cells and nerve cells of the esophagus are derived from small cells of the Sl cell group while the radial muscle cells are derived from the St cell group

The esophago-intestinal valve is formed from the same general tissues as the esophagus In the early postgastrular stage (Fig 152 BB) five nuclei may be seen between the esophagus and the intestine, at hatching (Fig 153 A) these five nuclei comprise a large dorsal nucleus, two subdorsal, one left lateral, and one ventral

The intestine in the early postgastrular stage (Fig 152 BB) consists of 2 lateral rows of 8 large cells derived from E With elongation there is a slight division of these cells and the two rows separate in the middle forming an irregular zigzag lumen surrounded by a dorsal and a ventral cell row

The rectum, in so far as known, is derived from the Sl cell group This group being entirely enclosed at gastrulation The proctodaeum is formed (Fig 152 DD) through the separation of cells in this region A group of 11 small cells lies between the posterior end of the intestine and the ventral side of the body As in the case of the esophagus the nuclei later separate through elongation of the organism Four cells surround the proctodeum at its junction with the body wall (AG1 and AG2) two being dorsal and two ventral, two lateral cells are anterior to these (Tg) a group of three large cells one dorsal and two ventral (Dg) lies anterior to these, and there are two additional cells, one dorsal and one ventral, connecting the intestine and rectum No increase in number of cells takes place in later development

Soon after the completion of gastrulation the genital primordium is recognizable as four cells two of which (the terminal cells) cover the other two (the primordial germ cells) Martini considers the terminal cells as probably originating from the M cell group It seems more probable, in the light of Pai's observations on Turbatrix aceti (See p 220), that the anterior cell resulting from the fifth cleavage in the P cell line (so called P4 1 or S5) formed this layer In case Pai is correct the two primordial germ cells present at hatching resulted from the sixth cleavage of the P stem cell

Regarding the development of the nervous system little is known except that it may form the small cells of the Sl and Sr cell groups Nothing whatsoever is known regarding the origin of the excretory system

Parascaris equorum (Fig 154) The embryology of the horse ascarid usually called Ascaris megalocephala, has been worked on by Boveri (1892, 1899 1909, 1910 a, b) zur Strassen (1896, 1899 a, b), Muller (1903), Zoja (1896) Bonfig (1925) Grigoloff (1911) Hogue (1911), Schleip (1924) and Stevens (1925) and in most of the results there is entire agreement The lineage has been followed up to the 802 cell stage by Muller at which stage the embryo is completely developed and somewhat elongate but has not reached the first larval stage The large number of cells and the difficulty of following postembryonic stages due to the life history of the species makes it impractical to follow the differentiation of particular tissues

In the general embryology Parascaris equorum is nearly identical with Turbatrix aceti but Boveri's beautifully illustrated work (1899) shows that chromatin material is lost from the nuclei during the division of somatic stem cells a fact which indicates a very definite cytological basis for the potentiality of the embryonic blastomeres Chromatin diminution is not known in other groups of nemas though the same differentiations in potentialities are present

The cleavage pattern of Parascaris equorum (Fig 154) is identical with that of the previously described species At the 56 cell stage (sixth cleavage) the cells are as follows 32 of the Sl group, 4 St, 4 M, 4 E, 8 C, 2 D and 2 P Thereafter all of the cells except the C and P lines divide (seventh cleavage) forming a 102 cell stage at which time there is a well formed gastrula (Fig 151 AA-EE) the anterior lip of which is bordered by 8 stomodeal cells (St) while the posterior lip is bordered by 4 proctodeal cells (D) During this division some of the Sl cells divide unequally and to those which have been more carefully traced in subsequent divisions Mueller (1903) gave a simplified terminology, g and G corresponding to pairs of cells as the fifth cleavage (such as A II') [*]further divisions forming ga, gb then gar gal, gbr, and gbl others were similarly renamed ogr upr, opl uql, arl kbr etc These cells contribute a large part of the

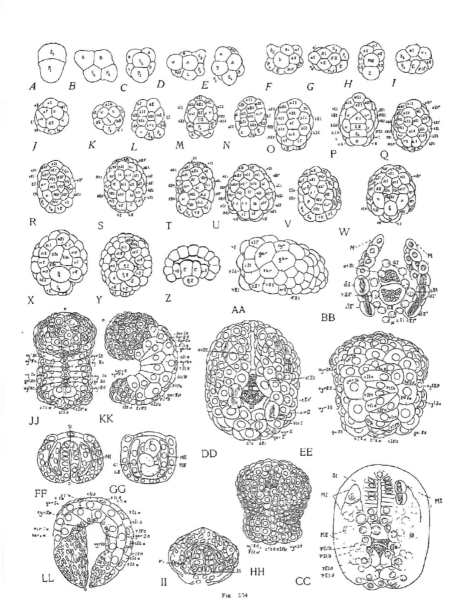

Fig 174

224

final body surface (Fig 154 JJ-KK) By the end of the seventh cleavage the eight large E cells nearly completely fill the blastocoele

At the eighth cleavage all cell groups with the exception of P4 divide so that a 202 cell embryo is formed its anterior surface is covered by small cells of the $S1$ group while much of the posterior part of the embryo is covered by the larger h, oy, uy, g and x cell groups (Fig 154 AA & EE) Cells of the C line ($c\ II'\ c\ II'$ etc) form paired posterior subdorsal rows of cells while those of the D line ($d\ II'\ D\ II$) enter the ventral groove The St cell group now consists of 16 cells, some of which extend as far posterior as the genital primordium (Fig 151 BB) while the remainder form an anterior groove in continuation with the primary ectoderm ($S1$) The M cell group consists of two irregular lateral groups of eight cells each, entirely enclosed as is the 16 cell E group by $S1$ and $S3$ cell groups The gastrular cavity is sharply V-shaped anteriad, lined by cells of the St group while it is U-shaped posteriad, the large genital primordium cells ($P4\ I$ and II) forming the ventral surface (Fig 154 CC) Parts of the $S3$ cell group ($c\ II$ and $c\ II$) are definitely mesodermal

At the ninth cleavage the embryo is 402-celled, being composed as follows $S1$ cell group 256, St group 32, M group 32, E group 32, $c\ I$ and $c\ I$ (Secondary ectoderm) group together 16, $c\ II$ and $c\ II$ (Tertiary mesoderm) group together 16, $S4$ (D) group 16, and $P4$ (G) group 2 The two subdorsal surface cell rows (Fig 154 FF-GG) are formed from $c\ I$ and $c\ I$, two large lateral mesodermal bands are formed from $M\ c\ II$ and $c\ II$ and part of St

The anterior part of the ventral groove has completely closed, the esophageal primordium forming a solid cell mass in contact with the ventral and anterior small cells of the primary ectoderm A terminal cavity the stomodeum is then formed in the esophageal primordium The most posterior St cells (Fig 154 BB) do not become a part of the esophageal primordium, though at this stage the maximum number of St cells should be 32 and though not all of them enter into the formation of the esophagus, there are about twice that number in the primordium The cells not accounted for were probably small $S1$ cells which entered the primordium during formation of the stomodeum

During the latter part of the ninth cleavage the dorsal cells of the C group come to form a single row of 10 very large cells covering the dorsal and posterior surfaces of the embryo, this being accomplished through median movement of alternate cells (Fig 154 JJ-KK) Anteriorly this row is continued by the cells designated $ka\ II\ B$, $kal\ II\ A$, their sister cells being lateral to them At the sides of this dorsal cell row there are 2 large subdorsal cell rows formed from the gai, gal, kal, kar, cell groups, and at the sides of these, 2 lateral large cell rows formed from the oy and uy cell groups These large cells are of particular significance for they swell in size and then cover most of the posterior and ventral cells of the $S1$ group, thus forming the epithelium of much of the body

At this time the embryo begins to elongate definitely, and becomes ventrally curved, this probably being due to swelling of the 5 large cell rows The lateral cell rows nearly come together, ventrally forcing some of the small superficial cells anteriad This is considered the completion of gastrulation The anterior end of the embryo and the ventral surface are covered by small cells of the $S1$ group We now find the mediodorsal and posterior medioventral parts of the embryo covered by cells derived from $S3$ ($c\ I$ and $c\ I$), the sides by cells of the $S1$ cell group (oy, uy, ka, kal, ga, gal, uy, and oy), and the

anterior and ventral part of the body by $S1$ A further division at least of the large surface cells takes place after elongation of the embryo into definite vermiform shape

Other species —The embryology of Rhabditis terricola, Diplogaster longicauda, and Nematoxys ornatus is, so far as known, similar to that of Rhabditis bufonis In Pseudalius minor no blastocoele is developed, the embryology being very similar to that of Camallanus lacustris In the case of Syphacia obvelata (Oxyuris obvelata) early cleavage is somewhat modified through the elongate "banana" form of the ovum The first cleavage is extremely, unequal, $S1$ being nearly twice as long as $P1$ This type of ovum is very common in oxyurids and thelastomatids The first cleavage of Metastrongylus elongatus appears equal but must be unequal since $P1$ contains a large amount of yolk material while $S1$ does not As in Camallanus, no blastocoele develops

Abnormal development Development is strongly determinate as would be indicated from the previous discussion Sometimes variations occur in the early cleavages, particularly in Parascaris Normal formation of rhomboid embryos in the four-cell stage is assured in most nematodes but in this form, due to the planes of the second cleavage, arrangement of the cells is observed which becomes rhomboid by passing through an]-shaped stage Sometimes however by passing through an [-shaped stage the positions of the blastomeres are reversed, B being anterior to A In such cases the entire development of the embryo is reversed, both A and B develop normally like B and A, the third somatic stem cell is formed at the opposite end of P, $S2$ divides normally and development proceeds to the blastula stage, development of Mst is probably influenced since gastrulation does not occur Injury of $P2$ in the 2-cell stage does not stop further development of the $S1$ and $S2$ cells up to the blastula stage, which is abnormal, injury through loss of cytoplasm in the $S1$ cell at the two-cell stage does not stop further development of the $P1$ cell in a normal manner The position of the spindle of the first cleavage may be changed through centrifuging or by multispermy, in either case the first cleavage may give rise to equipotential blastomeres which result in the formation of $P1$ and $S1$ cells, showing that the potentialities are dependent upon cytoplasmic material and that probably the occurrence of chromatin diminution in a blastomere is also dependent upon the cytoplasm Separation of $P1$ and $S1$ in Tubatrix aceti results in the degeneration of $S1$ while $P1$ continues development to the 4-cell, to 16-cell or gastrular stage

These observations appear to indicate that nemic embryos are essentially of mosaic structure, and that the unequal potentialities of the blastomeres are due to some differences in the cytoplasm but probably also to other factors such as influence from surrounding cells and differences in chromatin

Bibliography

AUERBACH, L 1874 —Organologische Studien zur Charakteristik und Lebensgeschichte der Zellkerne 262 pp, pls 1-4 Breslau

BONFIG, R 1925 —Die Determination der Hauptrichtungen des Embryos von Ascaris megalocephala Ztschr Wiss Zool v 124 (3-4) 407-456, figs 1-25

BOVERI, T 1892 —Ueber die Entstehung des Gegensatzes zwischen den Geschlechtszellen und den somatischen Zellen bei Ascaris megalocephala Sitz Gesellsch Morph & Physiol, v 8 (2-3) 114-125, figs 1-5

1899 —Die Entwickelung von Ascaris megalocephala mit besonderer Rücksicht auf die Kernverhältnisse Festchr Kupffer Jena 383-430, figs 1-6, pls 40-45, figs 1-45

1909 —Die Blastomerenkerne von Ascaris megalocephala und die Theorie der Chromosomenindividualität Arch Zellforsch, v 3 (1/2) 181-268, figs 1-7, pls 7-11 figs 1-51

1910a —Ueber die Teilung centrifugierter Eier von Ascaris megalocephala Festchr W Roux Arch Entwicklungsmech v 30 (2) 101-125, figs 1-32

1910b —Die Potenzen der Ascaris-Blastomeren bei abgeänderter Furchung Festchr R Hertwigs v 3 131-214, figs A-Y, pls 11-16, figs 1-39

Fig 154

Parascaris equorum A—2 cell stage B—4 cell C—6 cell D—8 cell E—8 cell F—10 cell G—12 cell, lateral view H—12 cell ventral view I—12 cell sagittal section J—16 cell K—16 cell L—22 cell M—24 cell ventral view N—24 cell lateral view O—26 cell dorsal view P—28 cell Q—41 cell dorsal view R—41 cell lateral view S—44 cell dorsal view T—48 cell dorsal view U—48 cell lateral view W—54 cell ventral view X 56 cell ventral view Y — 92 cell ventral view Z — 92 cell cross section AA — 202 cell lateral view (8th cleavage) BB 102 202 cell horizontal section (in 8th cleavage) CC EE—102 202 cell ventral ventral and dorsal views FF-GG —202 402 cell cross sections (in 9th cleavage), HH—402 802 cell ventral view II—Esophageal region of the embryo JJ KK—ventral and lateral views of same LL—Prelaval stage surface view

A W After zur Strassen 1896 Arch Entwickelungsmechanik v 3 (1 2) X Z After Boveri 1892, Sitz Gesellsch Morph & Physiol v 8 AA KK after H Mueller 1903 Zoologica (41)

BUTSCHLI O 1875a —Voiläufige Mittheilung über Untersuchungen betreffend die ersten Entwicklungsvorgänge im befruchteten Ei von Nematoden und Schnecken Ztschr Wiss Zool, v 25 (2) 201-213
1875b — Zur Entwicklungsgeschichte des *Cucullanus elegans* Zed Ztschr Wiss Zool, v 26 (7) 103-110, pl 5, figs 1 8

CONTE A 1902 — Contributions a l'embryologie des nematodes Ann Univ Lyon, n s, I Sc Med (8) 133 pp, figs 1 137

GAILB O 1878 —Recherches sui les entozoaires des insectes Organisation et developpement des oxyurides Arch Zool Exper & Gén, v 7 (2) 283-390, pls 17-26

GANIN M S 1877a —[Mittheilungen über die embryonale Entwickelung von *Pelodera teres*] (in Hoyer, H Protocolle der Sitzungen der Section für Zoologie und vergleichende Anatomie der V Versammlung russischer Naturforscher und Aerzte in Warschau im September 1876) Ztschr Wiss Zool, v 28 (3) 412-413
1877b —[Ueber die Untersuchungen von Natanson betreffend die embryonale Entwickelung von drei Arten von Oxyuris] Ztschr Wiss Zool, v 28 (3) 413-415

GIRGOLOFF S S 1911 — Kompressionsversuche am befruchteten Ei von *Ascaris megalocephala* Arch Mikrosk Anat v 76

GOETTE, A 1882 —Abhandlungen zur Entwickelungsgeschichte der Tiere Erstes Heft Untersuchungen zur Entwicklungsgeschichte der Würmer Beschreibender Teil 104 pp, 4 figs, pls 1-6 Hamburg u Leipzig

HALLEZ, P 1885 —Recherches sur l'embryologie et sur les conditions du développement de quelques nematodes 71 pp, 4 pls Paris

HOGUE M J 1910 —Ueber die Wirkung der Centrifugalkraft auf die Eier von *Ascaris megalocephala* Arch Entwicklungsmech, v 29 (1) 109-145, figs 1-42

JAMMES, L 1891 —Recherches sur l'organisation et le developpement des nematodes These 205 pp Paris

LIST, T 1893 —Zur Entwicklungsgeschichte von *Pseudalius inflexus* Duj Biol Centralbl v 13 (9/10) 312-313, 1 fig
1894 —Beitrage zur Entwicklungsgeschichte der Nematoden Diss 32 pp Jena

MARTINI E 1903 —Ueber Fürchung und Gastrulation bei *Cucullanus elegans* Zed Ztschr Wiss Zool, v 74 (4) 501-556 figs 1-8 pls 26-28 figs 1-35
1906 —Ueber Subcuticula und Seitenfelde: einiger Nematoden 1 Ztschr Wiss Zool, v 81 (4) 699-766, pls 31-33 figs 1-34
1907 —Idem II Ibid v 86 (1) 1-54, pls 1-3 figs 1-82
1908a —Die Konstanz histologischer Elemente bei Nematoden nach Abschluss der Entwicklungsperiode Anat Anz v 32 Erganz Heft Verhandl Anat Gesellsch 22 Vers 132-134
1908b —Ueber Subcuticula etc III (Mit Bemerkungen über determinierte Entwicklung) Ztschr Wiss Zool, v 91 (2) 191-235 13 figs
1909 —Ibid Vergleichend histologische Teil IV

Tatsachliches Teil V Zusammenhange und theoretische Betrachtungen Ztschr Wiss Zool, v 93 (4) 535-624, figs z-uu pls 25-26, figs 82-106
1923 —Die Zellkonstanz und ihre Beziehungen zu ardern zoologischen Vorwurfen Ztschr Anat & Entwick 1 Abt v 70 (1/3) 179-259

MÜLLER H 1903 —Beitrag zur Embryonalentwickelung der *Ascaris megalocephala* Zoologica Stuttg Heft 41, v 17 1-30, figs 1-12, pls 1-5, figs 1-24

NATANSON 1876 —Zur Entwickelungsgeschichte der Nematoden Arb der 5 Versammml russ Naturf u Aerzte Warschau Ztschr Wiss Zool, v 28 see Ganin, 1877b

NEUHAUS, C 1903 —Die postembryonale Entwickelung der *Rhabditis nigrovenosa* Jena Ztschr Naturw v 37, n F v 30 (4) 653-690, 1 fig, pls 30-32, figs 1-40

PAI, S 1927 —Lebenzyklus der *Anguillula aceti* Ehrbg Zool Anz v 74 (11-12) 257 270, figs 1-12
1928 —Die Phasen der Lebenscyclus der *Anguillula aceti* Ehrbg und ihre experimentell-mophologische Beeinflussung Ztschr Wiss Zool v 131 (2) 293-344, figs 1-80, tables 1-3

SCHLEIP, W 1924 —Die Herkunft der Polaritat des Eies von *Ascaris megalocephala* Arch Mikrosk Anat & Entwicklungsmech v 100 (3/4) 573-598, figs 1-17

SPEMANN H 1895 —Zur Entwicklung des *Strongylus paradoxus* Zool Jahrb Abt Anat, v 8 (3) 301-317, pls 19-21, figs 1-20

STEVENS, N M 1909 —The effect of ultra-violet light upon the developing eggs of *Ascaris megalocephala* Arch Entwicklungsmech, v 27 (4) 622-639, pls 19-21, figs 1-67

STRASSEN, O zur 1892 —*Bradynema rigidum* v S eb Ztschr Wiss Zool, v 54 (4) 655-747, pls 29 33, figs 1 98
1896 — Embryonalentwicklung der *Ascaris megalocephala* Arch Entwicklungsmech v 3 (1) 27-105 figs 1-24, pls 5-9, 49 figs (2) 133 190, figs 25-26
1903 —Geschichte der T-Riesen von *Ascaris megalocephala* Teil I Zoologica Stuttg v 17, Heft 40 1-37 figs A M pls 1-6, figs 1-67
1906 —Die Geschichte der T-Riesen von *Ascaris megalocephala* als Grundlage zu einer Entwicklungsmechanik dieser Spezies [Continuation of 1903] Zoologica, Stuttg v 17 (40) 89-342, figs N-YYYY

VOGEL, R 1925 —Zur Kenntnis der Fortpflanzung, Eireifung Befruchtung und Furchung von *Oxyuris obvelata* Bremser Zool Jahrb Abt Allg Zool v 42 (2) 243 271, figs A-V, pl 1, figs 1-6

WANDOLLECK, B 1892 —Zur Embryonalentwicklung der *Strongylus paradoxus* Arch Naturg 58 J, v 1 (2) 123-148, pl 9, figs 1-30

ZIEGLER H E 1895 —Untersuchungen über die ersten Entwicklungsvorgange der Nematoden Zugleich ein Beitrag zur Zellenlehre Ztschr Wiss Zool, v 60 (3) 351-410 pls 17-19

ZOJA R 1896 —Untersuchungen über die Entwicklung der *Ascaris megalocephala* Arch Mikrosk Anat, v 47 (2) 218-260, pls 13-14

M B CHITWOOD

Except for size, reproductive organs and related structures the majority of nemas are fully developed at the time of hatching The primitive nemas having no increase in cell number in most organs, undergo no gross morphological changes However some of the more highly specialized groups undergo changes in the character of the labial region, stoma and esophagus as well as changes in internal structure

There is no true metamorphosis in nemic development comparable to that occurring in insects since tissues are not destroyed and rebuilt Changes in gross body form are for the most part, of proportion rather than structure *Heterodera* is the most striking example of modified body form In this genus the first stage larvae are typical "eelworms" while the preadults are thickened and sac-like The female continues enlargement, assuming a pear-shape in the adult stage (Fig 115 N) while the male returns to the previous thread-like appearance (Fig 163 N)

Most of the developmental changes are *palingenetic* that is, features are derived from long evolution and concerned with adult existence However, many *cenogenetic* features occur which are purely larval adaptive features, interpolated into development to aid the larva in coping with its own separate existence Developmental changes may best be considered system by system

CUTICLE Ordinary postembryonic developmental changes of the cuticle are limited to the thickening of the cuticle and the development of such structures as caudal alae in the male Many cenogenetic features appear after hatching disappearing before or with the last moult, while the majority of palingenetic modifications appear with the last moult

The caudal alae and papillae of the male develop in the fourth stage larva when the tissues of the body draw away from the old cuticle and the adult structures are formed beneath it (Figs 156 S-T, AA & DD) The spines of adult *Hystrignathus* and cuticular plaques (Fig 23, SSS p 22) of adult *Gongylonema* (according to illustrations by Alicata 1935) appear in the fourth stage larvae while the collarettes of *Spinitectus* (Fig 23, p 22), spines of *Hystrichis* and the large trifid lateral alae of *Phyacephalus sexalatus* according to Seurat, 1913, first appear at the last moult Cuticular inflation around the head (Fig 23 RRR) of trichostrongyloids (*Longistriata hassalli*) first appears in the pre-adult stage Roberts (1934) reports that in the third stage larvae of *Ascaris lumbricoides* well developed lateral "membranes" are present, becoming very broad and fin like in the fourth stage diminishing in the fifth stage As has been previously mentioned (p 25), lateral alae are often much larger and wider in the larvae than in the adult, this being particularly true in the members of the Oxyuridae and Thelastomatidae It is interesting to note that such wide alae are not present when the larva is pressed from the egg shell Lateral alae probably function as "wings" or "immovable fins" to assist the larvae in locomotion and are no longer necessary once the nema has settled in a suitable place

The remaining cuticular modifications appear to be purely cenogenetic Wetzel (1931) described four large hooks placed in dorsolateral and ventrolateral positions in the fourth stage larvae of *Dermatoxys veligera* These hooks were lost at the subsequent moult

Change in size and shape of the tail is, perhaps, the most common post embryonic phenomenon In *Hystrignathus* the tail is distally obfurcate during the first three larval stages (Fig 155) while in *Strongyloides* it is digitate only in the infective larvae Buds appear on the tail of the third stage larvae of *Strongyloides* before its emergence from the cuticle of the second stage Among the Strongyloidea the occurrence of a long thin filiform tail in the second stage larva causes the infective larva to

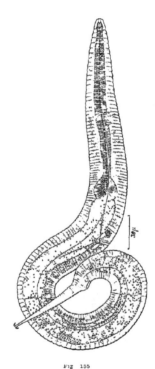

Fig 155

Hystrignathus rigidus, larval female showing forked tail After Christie 1934 Proc Helm Soc Wash v 1 (2)

have an even longer whip-like tail which is practically diagnostic for some groups (Fig 99), the tail of the third stage larva itself usually is quite short even conoid The tail of the second stage larva of trichostrongyles is shorter, conoid, attenuated with the tail of the third stage even more conoid inside it (*Trichostrongylus axei*, Fig 158 D) or it may bear papilla-like digitations, (*Ornithostrongylus quadriradiatus* Fig 158 U) which are subsequently lost while the pronged tail persists to the adult in *Ollulanus* and *Trichostrongylus*. In the Metastrongyloidea (Figs 156 U-V & 158 T) peculiar and characteristic notching of the tail under the cuticle is evident in the first or second stage becoming very well marked in the third stage and disappearing with the third moult

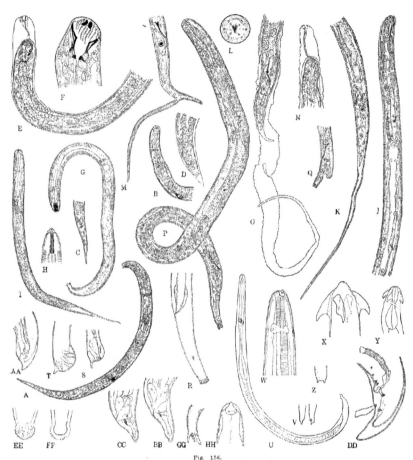

Fig. 156.

Postembryonic development. A-F—*Camallanus sweeti* [A—First stage larva. B—Anterior end, early second stage. C—Early second stage, posterior end. D-E—larva undergoing second moult, posterior and anterior end, respectively; F—Third stage larva, anterior end]. G-L—*Procamallanus pulchrocauda* (G—14 days old larva; H—Anterior end of larva, before first moult; I—Larva 6 days old. J-Q—*Dracunculus medinensis* (J—Anterior end of first stage larva; K—Posterior end of first stage larva; L—on face view of first stage larva; M—Posterior end of first stage larva; N—Cephalic region; O—Posterior end of normal larva undergoing first moult; P—Normal third stage larva; Q—Tail of abnormal third stage larva; R—*Tetrameres crami*, third stage larva. S-T—*Longistriata musculi* (S—Tail of preadult male; T—Tail of preadult male in final moult). U-V—*Aelurostrongylus abstrusus*, ? second stage larva. W—*Acanthocheilus rotundatus*, head ? third or fourth stage larva. X-Y—*Dermatoxys veligera*, head, fourth stage. Z—*Gongylonema pulchrum*, tail third stage larva. AA-BB—*Physocephalus sexalatus* (AA-Tail fourth stage male; BB-Tail of third stage larva). CC—*Ascarops strongylina*, tail third stage, DD—*Spirura gastrophila*, Tail male during fourth moult. EE-FF—*Seurocyrnea colini*, tail third stage. GG-HH—*Cheilospirura hamulosa*, third stage, tail and head. A-F, after Moorthy, 1938, J. Parasit., v. 24 (4). G-I, after Li, 1935, J. Parasit. v. 21 (2). J-Q, after Moorthy, 1938, Amer. J. Hyg. v. 27 (2). R, after Swales, 1936, Canad. J. Res. Sec. D., v. 14. S-T, after Schwartz & Alicata, 1935, J. Wash. Acad. Sci., v. 25 (5). U-V, after Cameron, 1927, J. Helminth., v. 5 (2). W, after Wuelker, in Wuelker and Stekhoven, 1933, Die Tierw. Nord-u. Ostsee, v. 5a. X, after Dikmans, 1931, Tr. Amer. Micr. Soc., v. 50 (4). Y, after Wetzel, 1931., J. Parasit. v. 18. Z-BB-CC, after Alicata, 1935, U. S. D. A. Tech. Bull. 489. AA, after Seurat, 1912, Compt. Rend. Soc. Biol., Paris, v. 75. DD, after Seurat, 1914. Ninth Congres internat. Zool. EE-HH, after Cram, 1931, U. S. D. A. Tech. Bull. 227.

Among many representatives of the Spiruroidea and Filarioidea the first stage larva bears hooks and spines in the cephalic region. These same forms sometimes have an attenuated tail (*Gongylonema pulchrum, Ascarops strongylina, Physocephalus sexalatus* and *Dichelonema rhea*). Both cephalic and caudal modifications are lost at the first moult. Since this is the stage of entry into the intermediate host these structures are probably used for boring through the tissues of the host. The occurrence of caudal specializations is characteristic of certain species of spiruroids. In the Thelaziidae, the tail is terminated by two or four small digitations in *Gongylonema pulchrum* (Fig. 156 Z) while in *Ascarops*, (156 CC) *Spirocerca* and *Physocephalus* (156 BB) it takes the form of a round knob, said knob being unarmed in *Ascarops strongylina* bearing a few spines in *Spirocerca lupi* and many rounded protuberances in *Physocephalus sexalatus*. According to Swales (1936) the tail of the third stage larva of *Tetrameres crami* (Fig. 156 R) is abruptly truncate bearing a circle of nine digitations of equal size and one subsequal median digitation; the tail of *Habronema muscae* has a rounded tip with many small spines while *Seurocyrnea colini* (Fig. 156 EE-FF) has a similar tail but the knob is larger and the spines relatively smaller. *Cheilospirura hamulosa* of the Acuariidae has a multi-pronged tail (Fig. 156 GG) in the third stage while in the same stage *Physaloptera turgida* has a bluntly conoid tail.

In the Camallanoidea and Dracunculoidea (Fig. 156) the first stage larva has a dorsal denticle (except *Micropleura*) on the head, [*]and an attenuated tail with large pocket-like phasmids. The dorsal denticle which is lost at the first moult, might be considered homologous to the hook present in the first stage *Gongylonema*. (Fig. 157 B-C). The tail of the third stage larva of *Dracunculus* has four large mucrones while that of *Camallanus* has three prongs, the tail of the adults being conoid and conically rounded respectively. However in *Procamallanus fulvidraconis* the three prongs persist to the adult.

The postembryonic changes in the tail of *Agamermis decaudata* Christie, (1936) are not limited to the cuticle. The division between the anterior and posterior portions of the body is marked by a node which is evident early in the preparasitic larva. The posterior part, about 2/5 of the entire length, becomes detached at the time of the entrance of the nema into the host.

HYPODERMIS. The discussion of postembryonic development in the hypodermis may be found on pp. 34-35 & 37. Briefly, it may be noted that in the more generalized nematodes little or no increase in number of cells or nuclei occurs after hatching while in more highly evolved forms many cell divisions or syncytial development may occur.

Teunissen (in Stekhoven, 1939) has found that the number of hypodermal glands in young individuals of *Anoplostoma granulosus* is from 46 to 70 in a quadrant or a total of 184 to 280. The number in juvenile males varies from 204 to 312 while in young females they number 244 to 356. Before sex can be determined, young specimens can be divided into two groups, those with 50 to 60 and those with 60 to 70 glands per quadrant. Cells increase in number as the nema grows, the number being constant only in the adult. The number of glands in females of 600 to 2000 microns length varies from 60 to 85 and the number in the male of the same size varies from 50 to 75. One can be nearly certain that young specimens with less than 55 hypodermal glands in a row will develop into males and that those having more than 65 cells will develop into females. Cell constancy in the individual is not reached until it is 1200 to 1400 microns in length (or adult) since the number of hypodermal glands is definitely larger in sexually mature specimens than in larvae.

MUSCULATURE. see p. 219.

NERVOUS SYSTEM. No changes known.

LABIAL REGION and STOMA. The labial region of most nemas is not as distinctly set off from the remainder of the body at the time of hatching as it is in the adult stage, though there are some notable exceptions to this

rule. Usually the cephalic papillae in parasitic nematodes are relatively larger and better developed at the time of hatching than in the adult. In rhabditids the labial region is distinctly set off and the amphids are oval, somewhat further posterior, and relatively larger than in the adult stage (Fig. 158 W-Z). In strongylins three or six indistinct lips are usually present in the first stage larvae though lips may be totally absent in the adult (*Ancylostoma caninum*). Furthermore, the cephalic papillae show a much more generalized pattern at this time, the internal circle and dorsodorsal and ventroventral papillae of the external circle being better developed than in the adult stage in which these papillae are greatly reduced. In ascaridids several changes take place in the labial structures. Young ascaridid larvae, broken out of the eggs have (*Ascaris lumbricoides* vide Alicata, 1935) three small lips bearing the full component of separate and well developed papillae (Fig. 57 X, p. 60) while the adult has large circumscribed lips with greatly reduced and partially fused papillae (Fig. 57 Y, p. 60). What

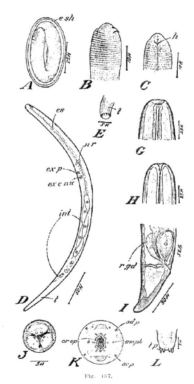

Fig. 157.

A-H—Developmental stages of *Gongylonema pulchrum*. (A—Fully developed larva in egg; B—First stage larvae, anterior end, lateral view; C—Same, ventral view, D—Larva, four days after experimental infection; E—Same, tail lateral view; G—Second stage larva, anterior end, lateral view; H—Same, dorsal view; I—Tail, lateral view). J—*Ascaris lumbricoides* larva from egg, en face K—*Physocephalus sexalatus*, third stage larva, en face. L—*Gongylonema pulchrum*, posterior end, ventral view. After Alicata, 1935, U. S. D. A. Tech. Bull. 489.

*Moorthy, 1938, described a dorsal appendage on some specimens of *Dracunculus medinensis*. In the first stage larva the appendage is long and filiform, while it persists throughout both second and third stages it is greatly reduced in size disappearing entirely at the third moult.

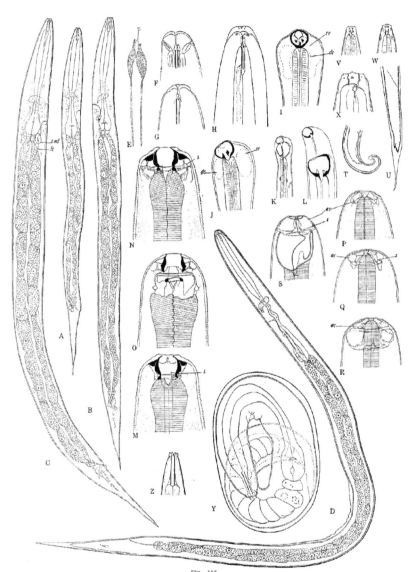

Fig. 158.

changes may take place in the intervening period we do not know, but in other members of the group there is some information Steiner (1921) described a larval ascarid (*Agamascaris odontocephala*) in which lips are absent, the cephalic papillae well developed, and in addition there is a mediodorsal tooth, or agamodontium (Fig 156 W) A similar tooth has been commonly observed in larval ascarids of fish (*Anacanthocheilus rotundatus* (Walker, 1930) Presumably this tooth aids in the migration of such larvae through tissues and is shed at the last moult

Spiruroids may pass through stages possibly indicating phylogenetic development but certain cenogenetic features sometimes occur Thus in the first stage larvae of *Gongylonema*, *Ascarops* and *Physocephalus*, Alicata has shown that there is a peculiar group of ventral cephalic hooks (Fig 157 C) and in the third stage larvae there are six indistinct rudiments of lips and all except the ventrolateral cephalic papillae are well developed Between the two dorsodorsal and the two ventroventral lips there is a pair of median cuticular elevations (Fig 157 K) Corresponding structures are present in the adult stage of *Simondsia* (Fig 58 R) However, all spirurids do not have the hooks and cuticular elevations In *Physaloptera* (Alicata 1937) the third stage larva has been found to have papillae and lips approximately as in the adult stage (Fig 58, p 63), unfortunately the first stage larva has not been studied The larva of *Habronema* (Hsu and Chow, 1938) has six lips instead of the adult two The first stage larva of *Camallanus* has an hexagonal oral opening and the labial region continues to be of larval form until the last moult at which two "lateral jaws" appear

Little is known of the development of labral structures in the Aphasmidia Crossman (1932) found that the larvae of *Tylocephalus* have a head resembling *Plectus* in that there are four large setose papillae and during development membranes or "cushions" form a web between these structures

Changes occurring during development of the stoma are often very marked in specialized nemas though little change takes place in the generalized forms In rhabditids the only noticeable change is in the diameter of the stoma which becomes wider with age, the absolute length may not change in postembryonic development Only the cheilorhabdions and prorhabdions are cast at moulting Related nemas such as diplogasterids may show some changes during development of the stoma As a rule, nematodes having short and proportionately wide stoma in the adult stage have a much more narrow stoma tending to be cylindrical or prismoidal in the larval stage (Fig 158 V-X) Thus in the development of strongyloids and trichostrongyloids the stoma in the first stage larvae is rhabditiform while in the second stage larvae it collapses and the cheilorhabdions and prorhabdions may simulate a stylet in the third stage (Fig 158 H) In the Strongyloidea there is a rather extensive reformation of stomata between the fourth and adult stages

The stoma of the fourth stage larva is usually rather short and wide and is termed a provisional buccal capsule In the late fourth stage a cavity is formed around this

structure Looss (1897) observed two such cavities in *Ancylostoma* one dorsal and one ventral, which gradually fused In *Strongylus* (Fig 158 P S) and *Cylicostomum* (Fig 158 M-O) Ihle and van Oordt (1923, 1924) observed a single anterior cavity (a-c) completely surrounding the provisional buccal capsule At the base of this a septum (b) is formed separating the anterior cavity and provisional buccal capsule from the remainder of the body Behind the septum a new cavity is formed outside the anterior end of the esophagus Around it the adult buccal capsule forms (Fig 158 N & R) The esophagus then is withdrawn and becomes attached to the base of the adult buccal capsule (Fig 158 O & S) The old lining of the esophagus is attached to the provisional buccal capsule In other strongylins the stoma may remain cylindrical to the adult stage (*Cylindropharynx*, Fig 55 C) Metastrongyloids differ from the foregoing in that the stoma is never rhabditiform so far as is known, mesorhabdions and telorhabdions are degenerate in the first stage In the later development of such forms the stoma may disappear (*Metastrongylus elongatus*) Young thelastomatids have a longer, more cylindrical stoma than adults and the same may be said of ransomnematids while in oxyurids remarkable changes have been described Ihle and van Oordt (1921) found that the larvae of *Oxyuris equi* have a massive pseudostom (Fig 97) formed by a dilation of the corpus the dilation being entirely absent in the adult One would judge this to be a purely cenogenetic feature related to feeding habits Larval subulurids have approximately the same type of stoma as do the adults while the larval ascarids, like the adults, have none In the Spiruroidea Chitwood and Wehr (1934) found that the stoma in some forms appears to go through stages which are known to occur in the adult of other forms Thus in the case of *Physocephalus* six cuticular projections of the prostom in the third stage larvae appear to form the lips of the adult stage while they retain their original larval position through development to the adult stage in *Ascarops* It has also been found that the stoma is more cylindrical in the third stage larva than in the adult Ransom (1913) describes the mouth cavity of the first stage larvae of *Habronema* as shallow becoming longer and cylindrical by the third stage Passing now to the Filarioidea we find that in some forms (*Dirofilaria immitis*) the third stage larva has a well developed cylindrical stoma while the adult has no distinct stoma However in the related genus *Litomosoides* the stoma in the Aphasmidia is in practically the same form as it is in the larvae of *Dirofilaria* Regarding stomatal changes in the Aphasmidia a little is known only in the cases of mermithids, trichuroids and dioctophymatoids Christie (1936) has found that the stoma of the embryo of *Agamermis* is represented by two small plates posterior to which there is a long narrow cuticular tube surrounded by esophageal tissue Within the esophageal tissue a large onchiostyl develops and gradually comes to surround and replace a part of the stoma or esophageal lumen at moulting In trichuroids, Fuelleborn (1923) described similar onchiostyls as developing in late embryonic stages and Lukasiak (1930) described a stylet in larval dioctophymatoids which had been removed from the egg shell

ESOPHAGUS Postembryonic changes in the esophagus of nemas are limited to parasitic forms, some changes occurring in nearly all the large parasitic groups Presumably the changes are correlated with the development of new feeding habits In general the earlier stages of the esophagus are more similar to that structure in *Rhabditis* than is the esophagus of the adult, but very little or no change takes place in the number of cells during development except in those nemas with reduplicate esophageal glands (see p 233) At the time of hatching thelastomatids usually have a esophagus consisting of a cylindrical corpus, isthmus and valvulated bulb but in a few genera of the Thelastomatidae (*Hammerschmidtiella Leidynema*, 4 others) the metacorpus becomes enlarged in the adult female Peculiarly, no such change in form takes place in the development of the males Because of the late appearance of the swelling it is not considered a nomologue of the swelling present in the rhabditoid esophagus Oxyurids usually have an esophagus like that of the adult during all stages of development (exception *Oxyuris equi* see p 78) No particular developmental changes have been noted in the esophagus of heterakids but ascaridids present many

Fig 158

Postembryonic development of members of the Rhabditina and Strongylina A-D—*Trichostrongylus* axei larvae [Trichostrongylidae] (A—First stage B—Early second stage C—Late second stage D—Third stage) E H—*Ancylostoma caninum* [Ancylostomatidae] (E — Third stage larva excretory apparatus F G H — Head F — dorsal view and G H — lateral view) (I — *Caigeria rchyscelis* [Ancylostomatidae] Head (I — Ventral view fourth stage J — Lateral view fourth stage K — Late fourth stage L—Moulting specimen) M O—*Cylicostomum* sp [Strongylidae] head of larva lateral view (M—Fourth stage N—Late fourth stage O—During fourth moult) P S—*Strongylus vulgaris* [strongylidae] fourth stage larval female anterior end (P R—Stages in formation of buccal capsule S—Moulting specimen) T—*Uncaria stylosa elongatus* [Metastrongylidae] (Posterior end of larva in a second moult) U—*Ornithostrongylus quadriradiatus* [Trichostrongylidae] third stage larva (Lateral view of tail) V X—*Pristionchus* sp. [Diplogasteridae] stomal region (V ? first stage larva lateral view W—Same specimen dorsal view X—Adult) Y Z—*Rhabditis strongyloides* [Rhabditidae] (Y—Embryo in egg shell Z—Stomatal region first stage larva) E H after Stekhoven 1927 Proc Roy Acad Amsterdam v 30 I L after Ortlepp 1937 Onderstepoort J v Sc v 8 (1) M O after Ihle and Oordt 1923 Ann Trop Med & Parasit v 17 (1) P S after Ihle and Oordt 1924 Koninklijke Aknd Wetensch Amsterdam v 27 (4) T after Schwartz and Alicata 1931 J Parasit v 28 U after Cuvillier 19 7 U S D A Tech Bull 530 Remainder original

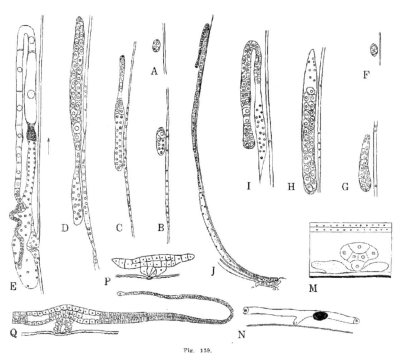

Fig. 159.

Postembryonic development of the reproductive system. A-J—
Turbatrix aceti [Diplogasteridae] (A—Genital primordium of newly
hatched female; B—24 hours; C—Three days; D—Five days; E—
Ovary of nine day old female; F—Genital primordium of newly
hatched male; G—Second day; H—Four days; I—Five days; J—
Testis; K-L—*Syphacia obvelata*, female reproductive system, im-
mature and adult. M-O *Ganguleterakis securatum*, genital primordium

(M—Third stage larval female; N—Late third stage female; O—
Fourth stage). P-Q—*Guigeria pachyscelis* genital primordium (P—
Fourth stage larval female. Q—Late fourth stage female repre-
ductive system, 3 weeks old). A-J, after Pai, 1928, Ztschr. Wiss.
Zool., v. 131 (2). K-L, after Vogel, 1925, Zool. Jahrb. Abt. Zool.
& Phys., v. 42. M-O, after Seurat, 1920, Hist. Nat. Nem. P-Q, after
Ortlepp, 1927, Onderstepoort J. Vet. Sci. v. 8 (1).

interesting variations. The first stage larvae of *Ascaris*,
Toxocara and *Toxascaris* all have an esophagus which
slightly resembles that of rhabditids or more precisely *An-
giostoma plethodontis*. It consists of a somewhat clavate
corpus, an indistinct isthmus, and a short pyriform bulbar
region. Information on the later development is lacking but
the adults have a cylindrical esophagus which in the case
of *Ascaris* and *Toxascaris* is not grossly subdivisible into
separate regions. In the case of *Toxocara* the posterior
end is set off as a muscular *ventriculus*, which apparently
corresponds to the reduced bulb of the first stage larva.
When an esophageal diverticulum is formed (*Contracaecum*), it develops as an evagination of the ventral side
of the bulbar region or ventriculus. The esophagi of
strongylins also pass through a very interesting series
of changes. In two superfamilies, the Strongyloidea and
Trichostrongyloidea the esophagus of the first stage
larva is usually identical with that of rhabditids (Fig.
158 A); during the second larval stage the valves degenerate (Fig. 158 C) and in the third stage the esophagus becomes long and narrow resembling the esophagus
of *Diplogaster* except that the swelling at the base of the

corpus is very indistinct (Fig. 158 D); in later development it becomes more or less clavate, obliterating nearly
all signs of former division. Similar changes take place
in the Metastrongyloidea except that the phylogenetic
reminiscence of rhabditoid affinities is not so marked
since even the esophagus of the first stage larva does
not have a valvulated bulb but resembles more closely
that of third stage larvae of strongyloids and metastrongyloids. Two families of the Rhabditoidea, the Rhabdiasidae
and the Strongyloididae, undergo change in the form of
the esophagus during development of the parasitic generation. In the first family the esophagus of the free-living
generation and of the first stage of the parasitic generation is rhabditiform while the later stages of development
of the parasitic generation show changes entirely comparable to the strongyloids. In the second family the
esophagus of the free-living generation and of the first
stage larvae of the parasitic generation is rhabditiphani-
form while in the later development of the parasitic
generation changes comparable to those of rhabdiasids
occur except that the esophagus of the adult remains much

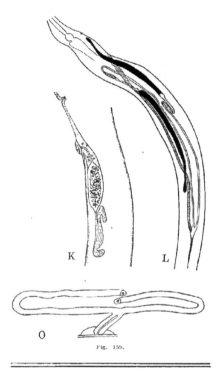

Fig. 159.

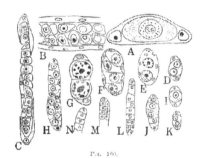

Fig. 160.

Genital primordia. A—*Bradynema ciguum*, B-C—*Rhabdias bufonis*, D-H—*Rhabditis aberrans*, I-J—*Allantonema mirabile*, K-L-N—*Bradynema strasseni*, M—*Allantonema mirabile*. All from Musso, 1930. Ziechr. Wiss. Zool. v. 137 (2). A, After Zur Strassen, 1892. B-C, after Neuhaus, 1903. D-H, after Krueger, 1913. M, after Wuelker, 1923.

as in the third stage strongyloid larvae, hence strongyli-form.

Little is known regarding the character of the esophagus of the first stage larvae of spirurins. In the first stage larvae of the genera *Ascarops* and *Gongylonema* Alicata (1935) found faint indications of a division into corpus, isthmus and bulbar region (Fig. 157 D) which entirely disappear during later stages and are replaced by a division into a short narrow anterior part and a long, wide posterior part, both parts being cylindrical. The first stage larvae of mermithids (*Agamermis decaudata*) may have an esophagus consisting of five regions; (1) and (2) equivalent to corpus, (3) a narrow part (? isthmus), (4) a swelling and (5) a long narrow posterior part. Two small subventral and a large dorsal esophageal gland are situated posterior to (4). In addition, two subventral rows of eight smaller cells, the stichocytes are situated along side the posterior narrow region (Fig. 93). During later development the esophagus narrows, the anterior swelling (2) disappears and the posterior part (5) becomes more or less surrounded by the large pare-sophageal body, the stichosome, which retains the two-cell-row form throughout later development. Each stich-ocyte is a large unicellular gland with a separate orifice. The three original esophageal glands atrophy after the nematode enters the host (See p. 92).

Very little is known about the esophagus of first stage *Trichuris* larvae but Fuelleborn (1923)

illustrated it as being composed of an anterior part terminated by a glandular swelling and a posterior part extending between two rows of stichocytes. According to Wehr (1939) the esophagus of the first stage larva of *Capillaria columbae* consists of a long narrow anterior part, a slight swelling and a cell body, or sticho-some region, consisting of a double row of seven sticho-cytes (Fig. 163 O). In the late first stage the stichosome has greatly increased in size and number of cells (Fig. 163 P). By the third stage the two rows have fused forming a single row of cells (Fig. 163 Q). The approxi-mate ratios of the length of the esophagus to the length of the intestine in each stage were given as follows: First stage 3. 5 : 1; Second stage 2 : 1; third stage 1. 8 : 1; fourth stage, 1. 1 : 1 and adult, 1 : 1.4.

Considering the esophagi of both trichuroids and mermithoids it seems reasonable to conclude the double row stichosome of first stage trichuroid larvae is palin-genetic. The intraesophageal character of the primary esophageal glands of trichuroids is undoubtedly primitive while their extraesophageal position in mermithids is recent and their hypertrophy at the period of penetration must be considered cenogenetic.

INTESTINE. Information on the postembryonic develop-ment of the intestine is strangely lacking. A few bold writers have admitted the presence of cells in the intestine but as a rule the intestine merely forms a connecting link between esophagus and rectum. However, changes both interesting and extensive do occur in some forms. In the more primitive nematode groups the multiplication of cells must be very limited and sometimes is probably confined to certain regions of the intestine. This appears to be true of most rhabditids, oxyuroids and similar forms. Numerous cell divisions must take place in the more highly evolved aphasmidian forms and in ascaridoids, spirurins, trichuroids and dioctophymatoids. The inte-stine of first stage *Ascaris lumbricoides* consists of innu-merable cells. According to Moorthy, 1938, the intestine of the late first stage larva of *Camallanus sweeti* has about 35 cells, the number increasing until in the early adult stage there are about 200. The same author records 12-15 cells in first stage *Dracunculus medinensis* while in the late third stage there are 36-40 cells. Of strongylids, known to possess few intestinal cells in the adult stage, Lucker (1935, 1936, 1938) offers considerable information. *Cylicodontophorus bicoronatus*, *Cylicocercus pateratus* and *Cylicocyclus insigne*, each have only eight intestinal cells in the infective larvae while the genera *Gyalocephalus* and *Strongylus* have 12 in the first genus and 16 to 32 in the latter. Lucker also definitely established the exist-ence of a lumen in these species with an eight cell intestine; according to his observations the intestine of second stage larvae of these forms have a 22 cell intestine, the number

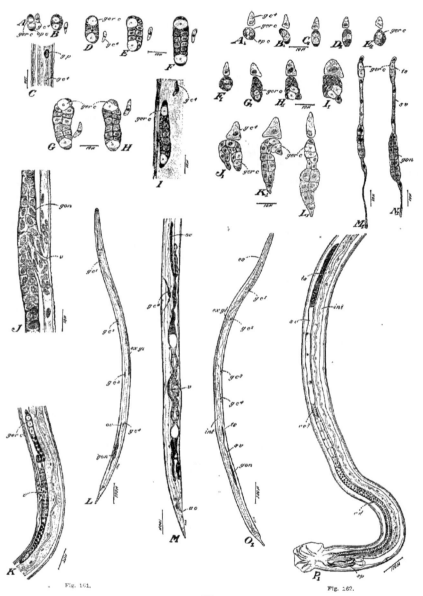

Fig. 161.

Fig. 162.

being reduced thereaftei Alcata (1935) records eight dorsal and eight ventral intestinal cells in both first and third stage larvae of *Hyostrongylus rubidus* The present writer found seven dorsal and seven ventral cells in the intestine of the first stage larvae of *Trichostrongylus axei*, 22 cells in the second stage and only 16 in the intestine of the infective third stage larvae The writer makes no attempt to explain the reduction in number of cells but the data were verified with numerous specimens Nuclear division without cell wall formation must occur later in the development of strongylins (Fig 102, p 102) In the case of mermithoids our information is more definite In many of these species cell division is followed by nuclear division without cell wall formation and finally fusion of syncytia and obliteration of the lumen may occur

Outpocketings or cecae have been previously described (p 100) in diverse groups of nematodes In ascaridins such cecae have been found to arise as evaginations of the intestinal epithelium during late larval development Similarly, Christie (1936) has found that the trophosome (the fat body or intestine) of mermithids grows anteriorly during larval development of *Agamermis decaudata* On this basis we may consider the trophosome in the esophageal region of mermithids as a caecum without a lumen

RECTUM, CLOACA and PERTAINING STRUCTURES In so far as is known, cell division does not take place in the postembryonic development of the posterior gut of the female It does, however, in the male for a small ventral growth of cells forms which is later joined by the vas deferens when it comes to open in the cloaca Similarly, there is a mass (or two masses) of cells from which the spicules develop The gubernaculum, on the other hand, is a cuticular thickening of the dorsal lining of the cloaca One may interpret the spicular sheaths as first an evagination of the dorsal wall of the cloaca, then an invagination of this structure (Fig 118 U) Both spicules and gubernaculum generally develop during the fourth stage

EXCRETORY SYSTEM Conclusive evidence is lacking with regard to the postembryonic development of the excretory system despite the numerous observations which have been made The primary cause of this failure is that all workers have proceeded on the assumption that a single ventral gland cell is the entire system Cobb (1890) described the excretory system of larval *Enterobius vermicularis* as a single invaginated cell from which the lateral canals and excretory vesicle developed In 1925 the same author described the first stage larva of *Rhabditis icosiensis* as similar to that of the adult except that the ventral gland was unpaired and the lateral canals free in the body cavity, the unpaired ventral cell was then supposed to divide forming the double glands of the adult The sinus and terminal duct nuclei were not accounted for Stekhoven (1927), Lucker (1935) and others have described a sinus (no nucleus seen) two subventral gland cells and no lateral canals in third stage larvae of the strongyloids The writer found the excretory pore, terminal duct, sinus, subventral gland cells and lateral canals all very plain in the first stage larva of *Trichostrongylus axei* (Fig 158 A-B) Before theorizing too much on the development of the excretory system, it would seem necessary that more critical data be obtained on the actual conditions existent in first stage larvae It seems possible that the so-called ventral gland or excretory cell usually described in larvae of parasitic nemas is actually the sinus cell and the terminal duct cell may be present but overlooked If this is the case, the system may originate from two germ lines in the Phasmidia The primary sinus nucleus might easily give rise to a secondary sinus nucleus and the paired subventral glands of the Strongylina and some rhabditids (*R icosiensis, R terricola, R strongyloides*) This would still not account for the lateral canals The theory that they develop from the sinus cell may be correct but it has not been demonstrated

FEMALE REPRODUCTIVE SYSTEM Development of the female reproductive system may be of two types, dependent upon the number of ovaries present in the adult In either instance the genital primordium of the first stage larvae consists of the same number of cells, four, arranged in the same manner as in the males *Turbatrix aceti* is the only one ovaried form that has been studied Pai (1928) found that after 24 hours the posterior somatic cell group (S5 II) has multiplied considerably, forming a mass of cells while the other cell groups (S5 I and P5) remained constant (Fig 159 B) Later all cell groups multiply (Fig 159 C-D) and the anterior end of the gonad bends posteriad while the posterior end (S5 II) grows posteriad also (Fig 159 E) The anterior somatic cell group forms the epithelium of the ovary while the posterior somatic cell group forms the oviduct, uterus, and seminal receptacle At this time an invagination of the hypodermis of the ventral chord meets the uterus forming the vagina and vulva

The gonad of *Tylenchinema oscinellae* was found by Goodey (1930) to develop in the same manner except that a twist occurs in the oviduct The vagina and uterus of *Sphaerularia* and *Atractonema* were found by Leuckart (1887) to become everted or prolapsed growing until the uterus is hundreds of times larger than the body in *Sphaerularia* (Fig 115A)

The development of the female reproductive system in nematodes with two ovaries (*Falcaustra lambdiensis Ganguria pachyscelis* and *Hyostrongylus rubidus*) differs in that both of the somatic cells form ovarian epithelium and both contribute to the formation of the uterus Division of the terminal cells results in an epithelial tissue covering the germinal cells with a terminal cell at the end of each ovary and a mass of somatic cells separating the two groups of cells resulting from the divisions of P5 I and P5 II (Fig 161) This mass of cells through further division and enlargement pushes the germinal cell groups apart and finally forms the uterus and oviducts At this time the middle part of the S5 group is joined to an invagination of the hypodermis forming the vulva and vagina, (Fig 161 J) Orlepp (1937) found that the ovejectors of *Ganguria pachyscelis* (Fig 159 P-Q) originate from the genital primordium and not the vulvar invagination Free-living nematodes with outstretched ovaries undergo no further development unless parts of the uteri are set off as seminal receptacles Free-living nematodes with opposed reflexed ovaries differ only in that the ends of the ovaries grow towards each other Scurat (1920) found that in the case of parallel ovaries or uteri in parasitic nematodes that the ovaries and uteri are at first outstretched, coiling and twisting of ovaries, uteri or both occur in very late larval or early adult development Another peculiarity in parasitic nematodes is that the uteri may become fused for part of their length forming either a continuation of the vagina (vagina uterina) or a common uterus The transformation from opposed to parallel oviducts and coincident development of a long uterine vagina was particularly well illustrated by Vogel (1925) in the development of *Syphacia obvelata* (Fig 159 K-L)

MALE REPRODUCTIVE SYSTEM The genital primordium of the male nematode consists of four cells at the time of hatching in all known cases Two of these cells the "terminal cells" cover the other two, the germinal cells Unfortunately the development has been traced only in nemas having a single testis in the adult Seurat (1918) discovered that the anterior end of the gonad of *Falcaustra lambdiensis* first grows anteriad, thereafter turning posteriad and extending to the cloaca, thus forming the vas deferens with the result that the gonad is flexed

Pai (1928) found that in *Turbatrix aceti* after 48 hours the genital primordium consisted of three groups of cells the primordial germ cells (P5), the anterior somatic cells (S5 I) forming a solid mass derived from the anterior terminal cell, and the posterior somatic cells (S5 II)

Fig 161

Postembryonic development of female reproductive system of *Hyostrongylus rubidus*, with position of coelomocyte adjoining genital primordium A—First stage B—Third stage C—Preparasitic third stage larva D—Third stage larva recovered 2 days after experimental infection E H—Third stage 4 days after experimental infection I—5 days after experimental infection larva on verge of third moult J—Vulvar region showing differentiation of ovary and gonoduct at 9 days L—Female larva after 9 days M—Young adult female posterior end All after Alicata 1935, U S D A Tech Bull 489

Fig 162

Postembryonic development of male reproductive system of *Hyostrongylus rubidus* A—First stage larva B1—Later first stage larva C—Late first stage larva D1 D2—Second stage larva E1—Preparasitic larva 1 1—2 days after experimental infection H1 J1—4 days after experimental infection K1—5 days after infection L1—6 days after M1 N1—Fourth stage larva 9 and 11 days after O1—11 days P1—Young adult male posterior end All after Alicata 1935, U S D A Tech Bull 489

form ng a covering for the germinal cells and terminal cell (Fig 159 G) After 96 hours no change had taken place except multiplication of cells (1 g 159 H) but after 120 hours the anterior somatic cells had grown posteriorly drawing the anterior part of the gonad with them (Fig 159 I) Thus the flexure of the testis takes place Finally the anterior somatic cells grow posteriorly and join the rectum forming the vas deferens and cloaca (Fig 159 J) Alicata (1935) found the development of the male reproductive system of Hyostrongylus rubidus to be essentially similar except that multiplication of germinal cells (P5) is delayed As in Turcatia, the anterior group of somatic cells (S5 I) bend posteriorly forming the vas deferens and seminal vesicle but the germinal zone is shifted around so that it is anterior and the gonad consequently is not flexed (Fig 162) Goodey (1930) on the contrary found that in Tylenchinema oscinellae the posterior terminal cell group extends posteriorly forming the vas deferens and joining with the rectum In this species the testis is not flexed

MICROFILARIA AND FILARIAL DEVELOPMENT Work on microfilaria has been developed as a separate science with little or no relationship to general nematology Since the pioneers were chiefly interested in identification of forms found in the blood of various species they developed a separate nomenclature for parts of the body Recent workers have made rapid strides in the identification of the parts of microfilaria with other nematodes

Some filariodes give birth to well formed first stage larvae or deposit well formed eggs containing such larvae These were placed in the family Filariidae by Wehr (1935) The larvae of such forms often have the cephalic hook and transverse rows of spines as seen in Gongylonema pulchrum (Fig 157 B), some of these have attenuated tails others rounded and spinate tails as in Gongylonema They are moderately well differentiated first stage larvae and in many cases, at least, should not be called microfilaria This term should be reserved for the rather unformed or embryonic young produced by the genera placed by Wehr in the Filariidae In these, stoma, esophagus, intestine and other organs are not completely differentiated

Microfilaria may be classified as to presence or absence of a sheath (Fig 163 J-k) The sheath is a very delicate membrane surrounding the larva, which some authors have considered a cuticle, indicative of the first moult, others a modified egg shell The fact that the sheath asserts chemicals in which a vitelline membrane would be dissolved, eliminates that possibility Evidence that the sheath develops from an egg membrane was presented by Penel (1904) and Seurat (1917) in the cases of Loa loa and Thamugadia hyalina (Fig 163 B-G) Its insolubility in alcohol and oils would signify that if it is an egg membrane, it is the shell There seems to be no morphologic difference correlated with presence and absence of a sheath

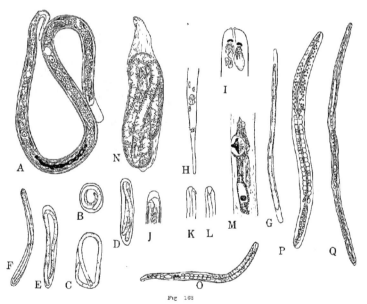

Fig 163

Postembryonic development continued A B E and N—Microfilaria loa (A—Entire larva H—Tail showing rhabdids I—Head M—Excretory pore and cell) B-G—Thamugadia hyalina successive stages J—Microfilaria bancrofti head K L—W recondita head lateral and ventral views N—Hartrounia sacchari fourth stage male in preparation for moult O Q—Capillaria columbae A H M after Fuelleborn 1929 lin Abuchn Path Micro v 6 28) B C after Seurat 1920 H st Nat Sem N after Strubell 1888 Bibliо Zool Org Abh Cesamut Zool v 2 O Q after Wehr, 1930 U S D A Tcl Bull 679

236

General Anatomy of Microfilaria For a comprehensive survey of the comparative anatomy of microfilarial species the reader is referred to Fuelleborn (1929) Regardless of the presence or absence of a sheath the embryo is covered by a delicate cuticle with distinct but minute striae (Rodenwaldt, 1908) The anterior end may bear a delicate hook (Fig 163 K) much like that of the first stage larvae of spiruroids like Gongylonema and filariids like Dicheilonema There is apparently no oral opening and the stoma, at most, is indicated by a primordium or vacuole. Paired "Mundgebilde" and 'Schwanzgebilde" have been demonstrated by intravitam stains (Fig 163 H-I), these seem to correspond to the amphids and phasmids

The remainder of the internal anatomy appears to consist of a rather disorganized mass of cells or nuclei some of which were sufficiently outstanding to have received specific names Manson (1908) named a clear spot with a massive adjoining cell situated in the cervical region, the V spot This was identified by Looss (1914) as the excretory pore and excretory cell (Fig 163 A & M) Rodenwaldt (1908) named four other large densely staining cells G1-4 considering them as genital cells and the vesicle with which they are associated, he termed GP or the genital pore Looss considered only the anterior-most of these (G1) as a genital cell and the remainder he named Z1-3 associating them with the rectum and correctly identifying the vesicle as proctodeum More recently Yamada (1927) and Feng (1936) have found that all four G cells take part in the development of the rectum (Fig 164 5-9) Beneath the cuticle there are four rows of spindle-shaped cells with elongate nuclei which Rodenwaldt named the matrix cells of the subcuticle According to more recent observations these are somatic muscle cells and total 14 to 62 in various species The remaining nuclei of the body constitute the nuclear column These take part in the formation of the esophagus, intestine nervous system and chords In the cervical region they are particularly numerous, interrupted only by the clear area indicative of the nerve ring In some species of microfilaria there is a clear area beginning some distance posterior to the excretory cell and ending anterior to G1, this structure termed the "Innenkorper" appears to be a yolk rest and corresponds to the lumen of the future intestine (Fig 163 A) Near the tip of the tail, there is sometimes a pair of tail nuclei which are lost at the first moult according to Feng

Later development For the later development of microfilariae, we will use as an example M malayi which has been very nicely worked out by Feng (1936) as far as the third stage (Fig 164) In this species the sheath is cast in the stomach of the mosquito on the first day Thereafter the organism is termed a first stage larva

until the fourth day at which the first moult occurs in the body of the mosquito The second stage is terminated by a moult on the sixth day and the third stage continues until after emergence from the intermediate host and entry to the final host During this period the nema grows from a length of 210 microns and diameter of 4 8 microns to a length of 1 3 mm and a diameter of 20 microns Growth is not uniform, after entry into the mosquito, there is first a shortening and widening, chiefly in the mid-region, the minimum length occurring after one and a half days (Fig 164 7) On the second day the mid-region starts to grow and after five and a half days, the esophagus begins suddenly to increase in length The width increases up to five and a half days, then diminishes The tail of the first stage larva is attenuated, that of the second stage short and conical and that of the third stage truncate, with two ventral and dorsal "papillae" (The ventral "papillae" are probably phasmids, while the dorsal is probably the tip of the tail) In Dirofilaria immitis two massive protuberant phasmids have been observed and no dorsal "papilla" exists The stoma begins to form on the second day (Fig 164 7) and by the fourth day (second stage) is indicated by refractile rods (protorhabdions), (Fig 164 9) In the third stage the stoma is conoid and the head bears the eight cephalic papillae and amphids characteristic of adult filariids

The esophagus first becomes distinct (Fig 164 8) on the third day and takes the typical two part glandular appearance of the adult by the fifth day The intestinal primordium is distinctly visible after 21 hours (Fig 164 6) and is well formed on the third day (Fig 164 8), by the fifth day there is a distinct lumen and the intestine is several cells in circumference The rectal primordium, G1-4 begins with a division of G1 after 24 hours (Fig 164 6) and a division of G2-4 accompanied by a second division of G1 after 34 hours, the rectal cuticle completes the formation of the 10 cell rectum on the third day (Fig 164 8) The genital primordium is first apparent on the fifth day (Fig 164 9) and is composed of seven cells (Fig 164 12), apparently it originates from cells in the intestinal region but its earlier existence has not been traced In both the first and second embryos the lining of the esophagus and rectum are moulted

Yamada (1927) had previously demonstrated two moults after exsheathment for Wuchereria bancrofti while the same number have been demonstrated in the intermediate host of non-ensheathed microfilaria such as Dirofilaria immitis It seems clear that the ensheathed microfilaria is an embryo rather than a larva Even after exsheathment the first stage larva is still rather embryonic in character and hardly deserves the term larva before the organs are demonstrable, this being on the third day for Microfilaria malayi (Fig 164 8)

Bibliography

ALICATA, J E 1935 —The tail structure of the infective Strongyloides larvae J Parasit, v 21 (6) 450-451 1 fig
1935 (1936) — Early developmental stages of nematodes occurring in swine U S Dept Agric Tech Bull 489, 96 pp, 30 figs
1937 (1938) —Larval development of the spiruroid nematode, Physaloptera turgida in the cockroach, Blatella germanica Papers on Helminthology, 30 Year Jubileum, K J Skrjabin, pp 11-14 figs 1-13

BOVIEN, P 1932 —On a new nematode Scatonema wulkeri, gen et sp n parasitic in the body cavity of Scatopse fuscipes Meig (Diptera nematocera) Vidensk Medd Dansk, Naturh Foren, v 94 13-32 figs 1-7
1937 —Some types of association between nematodes and insects Vidensk Medd Dansk Naturh Foren, v 101 1-114, figs 1-31

CAMERON, T W M 1927 —Observations on the life history of Aelurostrongylus abstrusus (Railliet), the lungworm of the cat J Helminth, v 5 (2) 55-66, figs 1-2

CHITWOOD B G and WEHR, E E 1934 —The value of cephalic structures as characters in nematode classification, with special reference to the superfamily Spiruroidea Ztschr Parasit, v 7 (3) 273-335, figs 1-20, 1 pl

CHRISTIE, J R 1934 —The nematode genera Hystrignathus Leidy Lepidonema Cobb and Artigasia, n g (Thelastomatidae) Proc Helm Soc Wash, v 1 (2) 43-48, figs 15-17
1936 —Life history of Agamermis decaudata, a nematode parasite of grasshoppers and other insects J Agric Res v 52 (3) 161-198, figs 1-20
1937 —Mermis subnigrescens, a nematode parasite of grasshoppers J Agric Res v 55 (5) 353-364, figs 1-6

COBB N A 1890 —Oxyuris-larvae hatched in the human stomach under normal conditions Proc Linn Soc N S Wales 2 s v 5 168-185, 1 pl
1925 —Rhabditis cossensis J Parasit, v 11 (4) 219-220 figs A-B

CRAM, E B 1931 —Developmental stages of some nematodes of the Spiruroidea parasitic in poultry and game birds U S Dept Agric Tech Bull No 227, 27 pp figs 1-25 1 pl

CROSSMAN L 1933 —Preliminary observations on the life history and morphology of Tylocephalus bacillivorus n g, n sp a nematode related to the genus Wilsonema J Parasit v 23 106-107

CUVILLIER E 1937 — The nematode, Ornithostrongylus quadriradiatus, a parasite of the domesticated pigeon U S Dept Agric Tech Bull No 569, 36 pp, figs 1-6

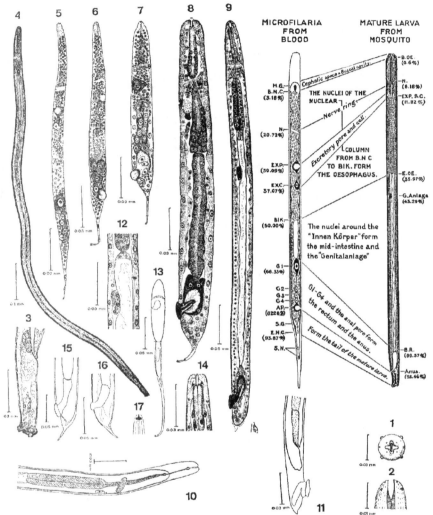

Fig. 164.

Development of *Microfilaria malayi* 1-15. 17—*Mf. malayi* (1-4—mature third stage larva, 1—en face; 2—lateral view of head; 3—tail, lateral view; 4—lateral view; 5—larva 2 hours after infection by mosquito; 6—24 hours in mosquito; 7—1½ days. 8—2½ days. 9—4½ days. 10-11—5½ days, anterior and posterior parts; 12—4½ days, showing genital primordium; 13—4½ days, abnormal development; 14—5½ days, showing buccal cavity and gland like structures; 15—4½ days, posterior end, lateral view). 16—*Mf. bancrofti*, 4½ days posterior end, lateral view. 17—*Mf. malayi*, mature larva, dorsal view. After Feng, 1936, Chinese Med. J. Suppl. 1.

238

DIKMANS, G 1931 —An interesting larval stage of *Dermatoxys veligera* Ti Amer Micr Soc, v 50 (4) 364-365, pl 29, figs 1-5

DIKMANS, G and ANDREWS, J S 1933 —A comparative morphological study of the infective larvae of the common nematodes parasitic in the alimentary tract of sheep Tr Amer Micr Soc, v 52 (1) 1-25, pls 1-6

DOBROVOLNY, C G and ACKERT, J E 1934 —The life history of *Leidynema appendiculata* (Leidy), a nematode of cockroaches Parasit, v 26 (4) 468-480, figs 1-10, pl 23, figs 1-3

ENICK, K 1938 —Ein Beitrag zur Physiologie und zum Wirt-Parasitverhaltnis von *Graphidium strigosum* (Trichostrongylidae, Nematoda) Ztschr Parasit 10 (3) 386-414, 1 fig

FENG, L C 1933 —A comparative study of the anatomy of *Microfilaria malayi* Brug, 1927 and *Mf bancrofti* Cobbold, 1877 Chinese Med J v 47 1214-1246, figs 1-6, pls 1-4
1936 —The development of *Microfilaria malayi* in *A hyrcanus* var *sinensis* Wied Chinese Med J, Suppl 1 345-367, pls 1-4
1937 (1938) —Studies on the development of Microfilariae Papers on Helminthology, 30 Year Jubileum, K J Skrjabin Moscow, pp 310-318 1 text fig, pl 1, figs 1-17

FULLEBORN F 1923 —Ueber den "Mundstachel" der Trichotracheliden Larven und Bemerkungen uber die jungsten Stadien von *Trichocephalus trichiurus* Arch Schiffs & Tropenhyg, v 27 421-425 pl 11, figs 1-18
1924 —Technic der Filarienuntersuchung Handb Mikr Technik, p> 2273-2304, figs 691-698, pls 13-14
1929 - Filariosen des Menschen Handbuch der pathogenen Mikroorganismen v 6 (28) 1043-1224, figs 1-77, pls 1 3

GOODEY, T 1930 —On a remarkable new nematode *Tylenchinema oscinellae* gen et sp n, parasitic in the Frit-fly, *Oscinella frit* L, attacking oats Trans R Soc Lond B v 218 315-343, pls 22-26, 1 fig

HSU, H F and CHOW, C Y 1938 —On the intermediate host and larva of *Habronema mansoni* Seurat, 1914 (Nematoda) Chinese Med J Suppl 2 419-422

IHLE, J E W and OORDT, G J VAN 1921 —On the larval development of *Oxyuris equi* (Schrank) Proc Sc K Akad Wetensch Amsterdam v 23 603-612 figs 1-6
1923 —On some strongylid larvae in the horse, especially those of *Cylicostomum* Ann Trop Med & Parasit, v 17 (1) 31-43, figs 1-9
1924 —Over de ontwikkeling van de larvae van het vierde stadium van *Strongylus edentatus* (Looss) koninklijke Akad Wetensch Amsterdam, v 33 (9) 865-872, figs 1-5
1924 —On the development of the larva of the fourth stage of *Strongylus vulgaris* (Looss) Koninklijke Akad Wetensch Amsterdam v 27 (3-4) 194-200, figs 1-5

LEUCKART, R 1887 —Neue Beitrage zur Kenntniss des Baues und der Lebensgeschichte der Nematoden Abhandl Math-Phys Cl K Sachs Gesellsch Wiss, v 13 (8) 565-704, pls 1-3

LI, H C 1935 —The taxonomy and early development of *Procamallanus fulvidraconis* n sp J Parasit, v 21 (2) 103-113, pls 1-2, figs 1-10

LOOSS, A 1897 —Notizen zur Helminthologie Egyptens 2 Centralbl Bakt 1 Abt, v 21 (24 25) 913-926, figs 1-10
1905 — Von Wurmern und Arthropoden hervorgerufene Erkrankungen Handb Tropenkrankh v 1 162 Vide Feng, 1936
1911 —The anatomy and life history of *Anchylostoma duodenale* Dub part 2 Rec Egypt Gov't School Med, v 4 159-613, pls 11-19, figs 101 208, photographs 1-41
1914 —Würmer und die von ihnen hervorgerufen Erkrankungen Handb Tropenkrank, 2 ed 433 vide Feng 1936

LUCKER, J T 1934 —The morphology and development of the preparasitic larvae of *Poteriostomum ratzii* J Wash Acad Sci, v 24 (7) 302-310, figs 1-12
1934 —Development of the swine nematode *Strongyloides ransomi* and the behavior of its infective larvae U S Dept Agric Tech Bull No 437, 30 pp, figs 1-5
1935 —The morphology and development of the infective larvae of *Cylicodontophorus ultraxectinus* (Ihle) J Parasit, v 21 (5) 381-385, figs 1-3
1936 —Comparative morphology and development of infective larvae of some horse strongyles Proc Helm Soc Wash, v 3 (1) 22-25, fig 9
1938 —Description and differentiation of infective larvae of three species of horse strongyles Proc Helm Soc, Wash, v 5 (1) 1-5, figs 1-2

LUKASIAK, J 1930 —Anatomische und Entwicklungsgeschichtliche Untersuchungen an *Dioctophyme renale* (Goeze, 1782) [*Eustrongylus gigas* Rud] Arch Biol Soc Sc & Let, Varsovic, v 3 (3) 1-100, pls 1-6, figs 1 30

MANSON, P 1903 —Tropical Diseases, a manual of the diseases of warm climates 756 pp, illus London

MARTINI, F 1920 —Bemerkungen zur Anatomie der Microfilarien Arch Schiffs & Tropen -Hyg, v 24 364-370

MOORTHY V N 1938a —Observations on the life history of *Camallanus sweeti* J Parasit, v 24 (4) 323 342, pls 1-4, figs 1-20
1938b —Observations on the development of *Dracunculus medinensis* larvae in Cyclops Amer J Hyg, v 27 (2) 437-460, pls 1-5, figs 1-19

MOORTHY, V N and SWEET, W C 1936 —A peculiar type of guinea-worm embryo Ind J Med Res, v 24 (2) 531-534, figs 1-6

MUSSO, R 1930 —Die Genitalrohren von *Ascaris lumbricoides* und *megalocephala* Ztschr Wiss Zool, v 137 (2) 247-363, figs 1-29, pls 1-2, figs 1-24

OIDIAM J N 1933 —On *Howardula phyllotretae* n sp, a nematode parasite of flea beetles (Chrysomelidae, Coleoptera) with some observations on its incidence J Helminth v 11 (3) 119-136, figs 1-3

ORTLEPP, R J 1923 —The life history of *Syngamus trachealis* (Montagu) von Siebold, the gapeworm of chickens J Helminth, v 1 119-140
1925 —Observations on the life history of *Triodontophorus tenuicollis* a nematode parasite of the horse J Helminth, v 3 1-14, figs 1-9
1937 —Observations on the morphology and life history of *Gaigeria pachyscelis* Rail and Henry, 1910 a hookworm parasite of sheep and goats Onderstepoort J Vet Sc, v 8 (1) 183-212, figs 1-18

PAI, S 1928 —Die Phasen des Lebenscyclus der *Anguillula aceti* Ehrbg und ihre experimentell-morphologische Beeinflussung Zts hr Wiss Zool, v 131 (2) 293-344, figs 1 80

PENEL 1905 —Les filaires du sang de l'homme Paris

RANSOM, B H 1913 —The life history of *Habronema muscae* (Carter), a parasite of the horse transmitted by the house fly U S Dept Agric B A I Bull No 163, 36 pp figs 1-41

RANSOM, B H and FOSTER, W D 1920 —Observations on the life history of *Ascaris lumbricoides* U S Dept Agric Bull No 817, 47 pp, figs 1 6

ROBERTS, F H S 1934 —The large roundworm of pigs, *Ascaris lumbricoides* L, 1758, its life history in Queensland, economic importance and control Bull No 1 Animal Health Station Yeerongpilly, pp 1-81, 11 figs, 2 pls

RODENWALDT, E 1908 —Die Verteilung der Mikrofilarien im Korper und die Ursachen des Turnus bei *Mf noctuna* und *diurna* Studien zur Morphologie der Mikrofilarien Arch Schiffs & Tropenhyg v 12 (10) 18 90
1933 —Zur Morphologie von *Microfilaria malayi* Meded Dienst Volksgezondh Ned -Ind Batavia, v 22 54-60, figs 1-2

SAIBAWA 1913—Untersuchungen uber Hundefilarien Centralbl Bakt Orig, v 67 68-75, 2 pls 1 text fig

SCHWARTZ, B and ALICATA, J E 1931—Concerning the life history of lung-worms of swine J Parasit, v 28 21-27, pl 1, figs 1-8
1935—Life history of *Longistriata musculi*, a nematode parasitic in mice J Wash Acad Sc, v 25 (3) 128-146, figs 1-14

SEURAT, L G 1913—Sur l'evolution du *Physocephalus sexalatus* Compt Rend Soc Biol Paris, v 75 517-520, figs 1-4
1914—Sur l'evolution des nematodes parasites 9th Congrès Internat Zool, Monaco pp 623-643, illus
1915—Sur les premiers stades evolutifs des spiropteres Compt Rend Soc Biol, Paris, v 78 561-564, figs 1-5
1916—Contributions a l etude des formes larvaires des nematodes parasites heteroxenes Bull Sci France & Belg, s 7, v 49 297-377, illus
1917—Filaires des reptiles et des Batraciens Bull Soc Hist Nat Afr Nord v 8 236
1919 Contributions nouvelles a l'etudes des formes larvaires des nematodes parasites heteroxenes Bull Biol France et Belg (1918) 52 344-378, illus
1920—Histoire naturelle des nematodes de la Berberie Première partie Morphologie, developpement ethologie et affinités des nematodes 221 pp, 34 figs Alger

STEINER G 1924—Some nemas from the alimentary tract of the Carolina tree frog (*Hyla carolinensis* Pennant) J Parasit v 11 1-32, pls 1-11 figs 1-65

STEKHOVEN J H S 1927—The nemas *Ancylostoma* and *Necator* Proc Roy Acad Amsterdam v 30 113-124, figs A-B, pls 1-3

SWALES, W E 1936.—*Tetrameres crami* Swales, 1933, a nematode parasite of ducks in Canada Morphological and biological studies Canad J Res Sec D, v 14 151-164, figs 1-10 1 pl

TEUNISSEN, 1939—In Stekhoven (1939) Nematodes Bronn's Klassen und Ordnungen des Tierreichs, v 4, Abt 2 Buch 3, Lieferung 6, pp 499-511, figs 94-98

VEVERS, G M 1921—On some developmental stages of *Ancylostoma ceylanicum* Looss, 1911 Proc Roy Soc Med v 14 25-27

VOGEL, R 1925—Zur Kenntnis der Fortpflanzung. Eireifung Befruchtung und Furchung von *Oxyuris obvelata* Bremser Zool Jahrb, Abt Zool & Phys, v 42 243-270, Figs A-T, 1 pl

WEHR E E 1935—A revised classification of the nematode superfamily Filarioidae Proc Helm Soc Wash, v 2 (2) 84-88
1939—Studies on the development of the pigeon capillarid, *Capillaria columbae* U S Dept Agric Tech Bull 679, 19 pp 3 figs

WETZEL, R 1931—On the biology of the fourth stage larva of *Dermatoxys veligera* (Rudolphi 1819) Schneider 1866 an oxyurid parasitic in the hare J Parasit, v 18 40-43, figs A-B

WULKER, G 1929—Ueber Nematoden aus Nordseetieren, I, II in Zool Anz, v 87 & 88

WULKER, G and STEKHOVEN, J H S 1933—Nematoda Allgemeiner Teil Die Tierwelt der Nord- und Ostsee Teil 5a 64 pp, figs 1-65

YAMADA, S 1927—An experimental study on twenty four species of Japanese mosquitoes regarding their suitability as intermediate hosts for *Filaria bancrofti* Cobbold Sci Reports Govt Inst Inf Disease v 6 559-622, 3 pls

Corrections

We are indebted to Dr G L Graham for his assistance in compiling this table of errors

Page 3 column 1, line 27, records to read records
Page 57 column 1, line 3, insert comma after ventro-ventrals
Page 59, column 2, line 8 of Fig 56, *cygne* should be *cygni*
Page 61 column 2, line 23, post labial to read postlabial
Page 63, column 1, line 41, *communis* to read *communis*
Page 64, column 2, line 13, *Mesomermis* to read *Mesomermis*
Page 66, column 1, line 39, insert comma before *Seleucella*
Page 66, column 2, line 10, *Hoplolaimus* to read *Hoplolaimus*
Page 67, column 1, line 22, carona to read corona
Page 67, column 1, line 44, sagitally to read sagittally
Page 67, column 2, line 43, sagitally to read sagittally
Page 67, column 2, line 72, plate like to read plate-like
Page 68, column 1, line 7 of Fig 60, FF to read F
Page 70, column 1, Fig 61, line 6, insert F before *Bathbolaimus cobbi* (section)
Page 73, column 1, line 46, *Trichuris trichura* to read *Trichuris trichiura*
Page 78, column 2, line 8, (L 13-19) to read (L 13-18)
Page 80, column 2, line 1, concentrated to read concentrated
Page 83, column 2, line 3, *Desmidocerca* to read *Desmidocercella*
Page 86, column 1, line 17, *Siphonolaimus* to read *Siphonolaimus*
Page 91, column 2 line 20, Leptosomatids to read leptosomatids
Page 95, column 1, line 5, *Immunck* to read Imminck
Page 95, column 1, line 14, comissure to read commissure
Page 95, column 1 line 58, *Angusticacum* to read *Angusticaecum*
Page 96 column 2, lines 5 and 6 of Fig 97 P = Q and Q = P
Page 98, column 1, Biblio, Buyn, *Angusticoecum* to read *Angusticaecum*
Page 101, column 1, line 25, mantel to read mantle
Page 103, column 1 Table 1, *Agameris* to read *Agamermis*
Page 104, column 1, line 26, postive to read positive
Page 112, column 1, Biblio Lucker, prepnasitis to read preparasitic
Page 113, column 1 line 9 *Travassos* to read Travassos
Page 115, column 1, line 13 of Fig 108, insert Q- before Longitudinal
Page 121, column 2 Biblio, Leuckart, Menschlichen to read Menschlichen

Note —

In order to avoid burdensome use of scientific names, in parts I and II we have used vernacular names for members of groups based on the following system

    stem plus —atins for suborder
                —oids for superfamily
                —ids for family
                —ins for subfamily

These endings were based on the older system of endings for the scientific names Dr Steiner points out that this is inconsistent with our use of the new style endings for the proper names In these and following parts the vernacular name for members of a subfamily will be dropped Where standard vernacular names for groups of closely related genera are in use they may be substituted (strongyles, comesomes, mononchs dorylaims) Otherwise the final vowel or vowels will be dropped from the group name Examples

Rank	Scientific name	Vernacular
Suborder	Rhabditina	rhabditin
Superfamily	Rhabditoidea	rhabditoid
Family	Rhabditidae	rhabditid

The correct stem of *Ascaris* is Ascarid- not Ascar- so the group names are correctly Ascaridina, Ascaridoidea and Ascaridae This makes burdensome vernacular names and we have not been able to eliminate the shorter form from the publication

Abbreviations

Lightning Source UK Ltd.
Milton Keynes UK
UKHW020257100223
416720UK00002B/619